MONOGRAPHS ON STATISTICS AND APPLIED PROBABILITY

General Editors

D.R. Cox, V. Isham, N. Keiding, N. Reid, and H. Tong

1 Stochastic Population Models in Ecology and Epidemiology
M.S. Bartlett (1960)
2 Queues *D.R. Cox and W.L. Smith* (1961)
3 Monte Carlo Methods *J.M. Hammersley and D.C. Handscomb* (1964)
4 The Statistical Analysis of Series of Events *D.R. Cox and P.A.W. Lewis* (1966)
5 Population Genetics *W.J. Ewens* (1969)
6 Probability, Statistics and Time *M.S. Bartlett* (1975)
7 Statistical Inference *S.D. Silvey* (1975)
8 The Analysis of Contingency Tables *B.S. Everitt* (1977)
9 Multivariate Analysis in Behavioural Research *A.E. Maxwell* (1977)
10 Stochastic Abundance Models *S. Engen* (1978)
11 Some Basic Theory for Statistical Inference *E.J.G. Pitman* (1979)
12 Point Processes *D.R. Cox and V. Isham* (1980)
13 Identification of Outliers *D.M. Hawkins* (1980)
14 Optimal Design *S.D. Silvey* (1980)
15 Finite Mixture Distributions *B.S. Everitt and D.J. Hand* (1981)
16 Classification *A.D. Gordon* (1981)
17 Distribution-free Statistical Methods, 2nd edition *J.S. Maritz* (1995)
18 Residuals and Influence in Regression *R.D. Cook and S. Weisberg* (1982)
19 Applications of Queueing Theory, 2nd edition *G.F. Newell* (1982)
20 Risk Theory, 3rd edition *R.E. Beard, T. Pentikainen and E. Pesonen* (1984)
21 Analysis of Survival Data *D.R. Cox and D. Oakes* (1984)
22 An Introduction to Latent Variable Models *B.S. Everitt* (1984)
23 Bandit Problems *D.A. Berry and B. Fristedt* (1985)
24 Stochastic Modelling and Control *M.H.A. Davis and R. Vinter* (1985)
25 The Statistical Analysis of Compositional Data *J. Aitchison* (1986)
26 Density Estimation for Statistics and Data Analysis
B.W. Silverman (1986)
27 Regression Analysis with Applications *G.B. Wetherill* (1986)

28 Sequential Methods in Statistics, 3rd edition
   G.B. Wetherill and K.D. Glazebrook (1986)
29 Tensor Methods in Statistics *P. McCullagh* (1987)
30 Transformation and Weighting in Regression *R.J. Carroll and
   D. Ruppert* (1988)
31 Asymptotic Techniques for Use in Statistics *O.E. Barndorff-Nielsen
   and D.R. Cox* (1989)
32 Analysis of Binary Data, 2nd edition *D.R. Cox and E.J. Snell* (1989)
33 Analysis of Infectious Disease Data *N.G. Becker* (1989)
34 Design and Analysis of Cross-Over Trials *B. Jones and
   M.G. Kenward* (1989)
35 Empirical Bayes Methods, 2nd edition *J.S. Maritz and T. Lwin* (1989)
36 Symmetric Multivariate and Related Distributions *K.-T. Fang
   S. Kotz and K.W. Ng* (1990)
37 Generalized Linear Models, 2nd edition *P. McCullagh and
   J.A. Nelder* (1989)
38 Cyclic and Computer Generated Designs, 2nd edition
   J.A. John and E.R. Williams (1995)
39 Analog Estimation Methods in Econometrics *C.F. Manski* (1988)
40 Subset Selection in Regression *A.J. Miller* (1990)
41 Analysis of Repeated Measures *M.J. Crowder and D.J. Hand* (1990)
42 Statistical Reasoning with Imprecise Probabilities *P. Walley* (1991)
43 Generalized Additive Models *T.J. Hastie and R.J. Tibshirani* (1990)
44 Inspection Errors for Attributes in Quality Control
   N.L. Johnson, S. Kotz and X. Wu (1991)
45 The Analysis of Contingency Tables, 2nd edition *B.S. Everitt* (1992)
46 The Analysis of Quantal Response Data *B.J.T. Morgan* (1993)
47 Longitudinal Data with Serial Correlation: A State-space Approach
   R.H. Jones (1993)
48 Differential Geometry and Statistics *M.K. Murray and J.W. Rice* (1993)
49 Markov Models and Optimization *M.H.A. Davis* (1993)
50 Networks and Chaos - Statistical and Probabilistic Aspects
   O.E. Barndorff-Nielsen, J.L. Jensen and W.S. Kendall (1993)
51 Number-theoretic Methods in Statistics *K.-T. Fang and Y. Wang* (1994)
52 Inference and Asymptotics *O.E. Barndorff-Nielsen and D.R. Cox* (1994)
53 Practical Risk Theory for Actuaries *C.D. Daykin, T. Pentikäinen
   and M. Pesonen* (1994)
54 Biplots *J.C. Gower and D.J. Hand* (1996)
55 Predictive Inference - An Introduction *S. Geisser* (1993)
56 Model-Free Curve Estimation *M.E. Tarter and M.D. Lock* (1993)
57 An Introduction to the Bootstrap *B. Efron and R.J. Tibshirani* (1993)

58 Nonparametric Regression and Generalized Linear Models
   P.J. Green and B.W. Silverman (1994)
59 Multidimensional Scaling *T.F. Cox and M.A.A. Cox* (1994)
60 Kernel Smoothing *M.P. Wand and M.C. Jones* (1995)
61 Statistics for Long Memory Processes *J. Beran* (1995)
62 Nonlinear Models for Repeated Measurement Data *M. Davidian and D.M. Giltinan* (1995)
63 Measurement Error in Nonlinear Models *R.J. Carroll, D. Ruppert and L.A. Stefanski* (1995)
64 Analyzing and Modeling Rank Data *J.I. Marden* (1995)
65 Time Series Models - In econometrics, finance and other fields *D.R Cox, D.V. Hinkley and O.E. Barndorff-Nielsen* (1996)
66 Local Polynomial Modeling and its Applications *J. Fan and I. Gijbels* (1996)
67 Multivariate Dependencies - Models, analysis and interpretation *D.R. Cox and N. Wermuth* (1996)
68 Statistical Inference - Based on the likelihood *A. Azzalini* (1996)
69 Bayes and Empirical Bayes Methods for Data Analysis *B.P. Carlin and T.A. Louis* (1996)
70 Hidden Markov and Other Models for Discrete-valued Time Series *I.L. Macdonald and W. Zucchini* (1997)
71 Statistical Evidence - A likelihood paradigm *R. Royall* (1997)
72 Analysis of Incomplete Multivariate Data *J.L. Schafer* (1997)
73 Multivariate Models and Dependence Concepts *H.Joe* (1997)
74 Theory of Sample Surveys *M.E. Thompson* (1997)
75 Retrial Queues *G. Falin and J.G.C. Templeton* (1997)
76 Theory of Dispersion Models *B. Jørgensen* (1997)
77 Mixed Poisson Processes *J. Grandell* (1997)
78 Variance Components Estimation - Mixed models, methodologies and applications *P.S.R.S Rao* (1997)
79 Bayesian Methods for Finite Population Sampling *G. Meeden and M. Ghosh* (1997)

(Full details concerning this series are available from the Publishers).

The Theory of Dispersion Models

BENT JØRGENSEN

*Associate Professor,
University of British Columbia,
Canada*

CHAPMAN & HALL
London · Weinheim · New York · Tokyo · Melbourne · Madras

Published by Chapman & Hall, 2–6 Boundary Row, London SE1 8HN, UK

Chapman & Hall, 2–6 Boundary Row, London SE1 8HN, UK

Chapman & Hall GmbH, Pappelallee 3, 69469 Weinheim, Germany

Chapman & Hall USA, 115 Fifth Avenue, New York, NY 10003, USA

Chapman & Hall Japan, ITP-Japan, Kyowa Building, 3F, 2-2-1 Hirakawacho, Chiyoda-ku, Tokyo 102, Japan

Chapman & Hall Australia, 102 Dodds Street, South Melbourne, Victoria 3205, Australia

Chapman & Hall India, R. Seshadri, 32 Second Main Road, CIT East, Madras 600 035, India

First edition 1997

© 1997 Chapman & Hall

Printed in Great Britain by St Edmundsbury Press, Bury St Edmunds, Suffolk.

ISBN 0 412 99718 8

Apart from any fair dealing for the purposes of research or private study, or criticism or review, as permitted under the UK Copyright Designs and Patents Act, 1988, this publication may not be reproduced, stored, or transmitted, in any form or by any means, without the prior permission in writing of the publishers, or in the case of reprographic reproduction only in accordance with the terms of the licences issued by the Copyright Licensing Agency in the UK, or in accordance with the terms of licences issued by the appropriate Reproduction Rights Organization outside the UK. Enquiries concerning reproduction outside the terms stated here should be sent to the publishers at the London address printed on this page.
  The publisher makes no representation, express or implied, with regard to the accuracy of the information contained in this book and cannot accept any legal responsibility or liability for any errors or omissions that may be made.

A catalogue record for this book is available from the British Library

∞ Printed on permanent acid-free text paper, manufactured in accordance with ANSI/NISO Z39.48-1992 and ANSI/NISO Z39.48-1984 (Permanence of Paper).

Contents

Preface	xi

1 Introduction to dispersion models — 1
- 1.1 Prelude — 1
- 1.2 Definitions — 3
  - 1.2.1 Dispersion models — 3
  - 1.2.2 Comments on definitions — 8
- 1.3 Analysis of deviance for generalized linear models — 12
- 1.4 Examples — 13
  - 1.4.1 Discrete data — 14
  - 1.4.2 Location-dispersion models — 17
  - 1.4.3 Positive data — 18
  - 1.4.4 Positive data with zeros — 20
  - 1.4.5 Directions — 20
  - 1.4.6 Proportions — 22
- 1.5 Properties of dispersion models — 24
  - 1.5.1 Unit deviance — 24
  - 1.5.2 Transformations — 25
  - 1.5.3 Saddlepoint approximations — 26
  - 1.5.4 Convergence to normality — 29
- 1.6 Outline of the rest of the book — 30
- 1.7 Notes — 31
- 1.8 Exercises — 32

2 Natural exponential families — 37
- 2.1 Moment and cumulant generating functions — 37
  - 2.1.1 Definition and properties — 37
  - 2.1.2 Convexity — 38
  - 2.1.3 Moments and cumulants — 40
  - 2.1.4 Characteristic functions — 41

2.2	Natural exponential families		42
	2.2.1	Definition	42
	2.2.2	Moment generating functions	44
	2.2.3	Moments and cumulants	46
	2.2.4	Support and steepness	47
2.3	Variance function and deviance		48
	2.3.1	Uniqueness theorem for variance functions	50
	2.3.2	Examples	52
2.4	Convergence results for variance functions		54
	2.4.1	Convergence of variance functions	54
	2.4.2	Asymptotic behaviour of variance function	57
2.5	Notes		63
2.6	Exercises		64

3 Exponential dispersion models — 71

3.1	Definition and properties		71
	3.1.1	Additive and reproductive forms	71
	3.1.2	Continuous and discrete models	75
	3.1.3	Unit deviance	77
	3.1.4	Convergence results	78
3.2	Convolution and additive processes		80
	3.2.1	Convolution formula	80
	3.2.2	Infinite divisibility	82
	3.2.3	Additive processes	83
3.3	Examples		85
	3.3.1	Normal distribution	86
	3.3.2	Gamma distribution	88
	3.3.3	Poisson distribution	90
	3.3.4	Binomial distribution	92
	3.3.5	Negative binomial distribution	94
3.4	Quadratic variance functions		97
	3.4.1	Morris classification	98
	3.4.2	Generalized hyperbolic secant distribution	100
3.5	Saddlepoint approximation		103
	3.5.1	Continuous case	103
	3.5.2	Discrete case	106
3.6	Residuals and tail area approximations		108
	3.6.1	Residuals	108
	3.6.2	Tail area approximations	111
	3.6.3	Examples	113
3.7	Notes		119
3.8	Exercises		121

4 Tweedie models — 127
4.1 Characterization and properties — 127
4.1.1 Scale transformations — 127
4.1.2 Cumulant generating function — 130
4.1.3 Unit deviance — 133
4.2 Special cases — 134
4.2.1 Stable distributions — 135
4.2.2 Stable Tweedie models — 135
4.2.3 Inverse Gaussian distribution — 137
4.2.4 Compound Poisson models — 140
4.3 Tweedie additive process — 144
4.3.1 General case — 144
4.3.2 Compound Poisson process — 146
4.4 Tweedie convergence results — 146
4.4.1 Regularity of variance functions — 146
4.4.2 Tweedie convergence theorem — 148
4.4.3 Convergence of Tauber type — 151
4.4.4 Examples — 155
4.5 Exponential variance functions — 160
4.5.1 Translation of exponential dispersion models — 160
4.5.2 Convolution formula — 162
4.5.3 Stable distribution with index 1 — 163
4.5.4 Convergence results — 164
4.6 Tweedie-Poisson mixtures — 165
4.6.1 Poisson mixtures — 166
4.6.2 Tweedie case — 167
4.7 Notes — 170
4.8 Exercises — 171

5 Proper dispersion models — 175
5.1 General dispersion models — 175
5.1.1 Definitions — 175
5.1.2 Properties — 177
5.1.3 Barndorff-Nielsen's formula — 181
5.2 Construction of proper dispersion models — 184
5.2.1 Renormalized saddlepoint approximations — 184
5.2.2 Transformational dispersion models — 186
5.2.3 Reproductive exponential families — 188
5.2.4 Non-transformational dispersion models — 190
5.3 Examples — 190
5.3.1 Generalized inverse Gaussian distribution — 190
5.3.2 Leipnik's distribution — 195

	5.3.3	Simplex distributions	198
	5.3.4	Simplex-binomial mixtures	201
	5.3.5	Tweedie contractions	204
5.4	Studentization		205
	5.4.1	From 'normal' to 'Student'	205
	5.4.2	From 'Student' to 'normal'	208
	5.4.3	Exponentiation	209
	5.4.4	Examples	210
5.5	Notes		215
5.6	Exercises		215

References   219

Symbol index   226

Author index   229

Subject index   231

Preface

This book is an introduction to the theory of dispersion models. The main *raison d'être* for dispersion models is to serve as error distributions for generalized linear models, introduced more than two decades ago by Nelder and Wedderburn. Generalized linear models are now fairly mainstream, and have inspired new developments in other areas such as for example longitudinal data and time series analysis. It hence seems timely to present a detailed study of dispersion models and their use in generalized linear models. The present volume deals with the theoretical aspects of dispersion models as such. A second volume is planned, on the statistical analysis for generalized linear models based on dispersion model error distributions.

The class of error distributions for generalized linear models had a humble beginning as a handful of well-known exponential family distributions. It gradually became clear that the class of possible distributions was large and rich enough to become a research topic in its own right. Eventually, the name *dispersion models* emerged, and the two main subclasses were identified, namely exponential dispersion models and proper dispersion models. Exponential dispersion models formalize the exponential family type of error distribution proposed by Nelder and Wedderburn, while proper dispersion models include a number of further useful distributions. Together they cover a comprehensive range of data types, including continuous, discrete and mixed data, and angles and proportions.

The theory of dispersion models straddles both statistics and probability, and involves an encyclopaedic collection of tools, such as exponential families, asymptotic theory, stochastic processes, Tauber theory, infinite divisibility and stable distributions. In particular, the study of variance functions for natural exponential families involves a variety of tools from probability and analysis, and has branched out as a separate research topic.

The common theme that emerges from this diversity of methods and models is the use of the deviance and the variance function. These two items determine the main aspects of the shape of the distributions. The systematic use of these two functions outside exponential families provides a convenient unification of results for exponential and proper dispersion models. In generalized linear models, the deviance and the variance function play key roles in analysis of deviance and residual analysis, respectively. This allows a unified approach to distribution theory and statistical analysis for generalized linear models based on dispersion models.

Concerns about application of the theory to generalized linear models have shaped the exposition in various ways. I concentrate on the univariate case, which is the most important one from a practical point of view. I emphasize the interpretation of the distributions and the development of models for different data types, and emphasize models that are useful for statistical data analysis. In order to make the exposition more accessible to graduate students, I have included introductory material on moment generating functions and natural exponential families.

This approach should make the book useful as both a research reference and a graduate-level textbook in distribution theory for generalized linear models. Prerequisites include measure-theoretic probability and some background in statistical inference. However, measure theory is used mainly in the definitions of natural exponential families and exponential dispersion models, and most of the results are accessible without detailed knowledge of measure theory.

I want to thank the following for comments on earlier versions of the manuscript: Ole E. Barndorff-Nielsen, Billy Ching, Philip Hougaard, Célestin Kokonendji, Rodrigo Labouriau, Philippe Lambert, Steffen Lauritzen, Gérard Letac, Jim Lindsey, Søren Lundbye-Christensen, Raúl Martínez, Peter McCullagh, John A. Nelder, V. Seshadri, Gordon Smyth, Peter Xue-Kun Song, Peter Thyregod and Min Tsao. I owe special thanks to Leah Min Li for help with the bibliography, and to Peter Thyregod for producing most of the plots.

Vancouver, December 1996   Bent Jørgensen

CHAPTER 1

Introduction to dispersion models

The theory of errors was developed by Gauss primarily in relation to the needs of astronomers and surveyors, making rather accurate angular measurements. Because of this accuracy it was appropriate to develop the theory in relation to an infinite linear continuum, or, as multivariate errors came onto view, to a Euclidean space of the required dimensionality. The actual topological framework of such measurements, the surface of a sphere, is ignored in the theory as developed, with a certain gain in simplicity. It is, therefore, of some little mathematical interest to consider how the theory would have had to be developed if the observations under discussion had in fact involved errors so large that the actual topology had had to be taken into account. The question is not, however, entirely academic, for there are in nature vectors with such large natural dispersions.

From Fisher (1953): 'Dispersion on a Sphere'

1.1 Prelude

The normal distribution was for a long time the main workhorse of statistical data analysis, associated with a large body of familiar theory. While the normal distribution is useful for many types of data, Fisher's statements quoted above suggest that the normal distribution is the exception, rather than the rule, except for data with small dispersions. Fisher goes on to remind us of the importance of describing data in their natural habitat, be it the real line, the sphere, the integers, or whatever. Analysis of non-normal data should hence take into account the actual form of the sample space for each type of data. This is easy enough with the aid of a computer, although each case may require some new theory.

However, the usefulness of the normal distribution lies not only in the ease of the analysis, but also in the ease with which new problems may be handled by using well-known and trusted techniques, involving no new theory as such. A case-by-case approach to non-normal data analysis is undesirable because of the burden of both developing and disseminating new theory for each new case.

Nelder and Wedderburn (1972) were the first to show, by introducing the class of **generalized linear models**, that a large variety of non-normal data may be analysed by a simple general technique. Their method, called **analysis of deviance**, generalizes the traditional method of **analysis of variance** for normal data and incorporates well-known methods for loglinear models and logistic regression, thereby bridging the gap between discrete and continuous data. Generalized linear models were originally developed for exponential families of distributions, but the main ideas may be extended to a wider class of models, called **dispersion models**, which is the main topic of the present book.

The class of dispersion models covers a comprehensive range of non-normal distributions, including distributions for the following seven basic data types, where S denotes the support of the distribution:

- Data on the real line, $S = \mathbf{R}$.

- Positive data, $S = \mathbf{R}_+$.

- Positive data with zeros, $S = \mathbf{R}_0 = [0, \infty)$.

- Proportions, $S = (0, 1)$.

- Directions, $S = [0, 2\pi)$.

- Count data, $S = \mathbf{N}_0 = \{0, 1, 2, \ldots\}$.

- Binomial data, $S = \{0, \ldots, m\}$.

Figure 1.1 shows some typical histograms for simulated samples for each of these seven data types, directions being illustrated by points on a circle. For discrete data, the observed proportions are indicated by bars, and for positive data with zeros, the bar in zero indicates the proportion of zeros, while the histogram indicates the distribution of positive observations.

The main idea behind dispersion models is that the notions of location and scale may be generalized to **position** and **dispersion**, respectively, for all the above seven data types. Similarly, the residual sum of squares from analysis of variance may be generalized to the notion of **deviance**, making analysis of deviance available as a general inference tool for a wide range of data types.

DEFINITIONS 3

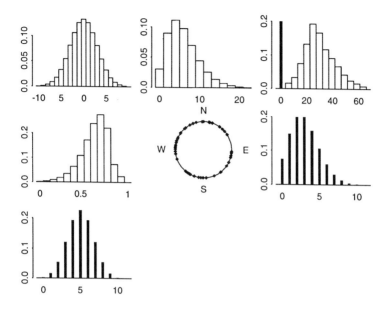

Figure 1.1. *Some basic data types illustrated by histograms of simulated data—clockwise from top left: Real, positive real, positive real with zeros, positive counts, binomial counts and proportions. Centre: directions.*

1.2 Definitions

1.2.1 Dispersion models

We now give the main definitions for dispersion models. By way of motivation, let us first consider the density function of the normal distribution $N(\mu, \sigma^2)$,

$$p(y; \mu, \sigma^2) = (2\pi\sigma^2)^{-\frac{1}{2}} \exp\left\{-\frac{1}{2\sigma^2}(y-\mu)^2\right\}.$$

From our point of view, the crucial properties of this density function are:

1. The exponent $-(y-\mu)^2/(2\sigma^2)$ is a negative constant times the squared distance between y and μ.

2. The factor $(2\pi\sigma^2)^{-1/2}$ does not depend on μ.

Another important property of the normal distribution is that μ and σ are location and scale parameters, respectively, but this is of secondary importance here.

The crucial step is to generalize the notion of squared distance, which we do in the next definition where we define the unit deviance. The unit deviance is particularly important from a statistical point of view, because sums of unit deviances are used in analysis of deviance in much the same way that sums of squares are used in analysis of variance.

In the following, we consider a family of distributions for a random variable Y. We let S denote the **set of realizable values** for the family, that is, the union of the supports of its members. The corresponding **convex support** C is defined as the smallest interval containing S. We let μ be a parameter with domain Ω, an interval contained in C. In most cases we have $\Omega = \operatorname{int} C$ (where int denotes 'the interior of'), but to accommodate certain special cases, we allow Ω to be different from $\operatorname{int} C$. In the following, the second derivatives of a function of two arguments, say $f(y;\mu)$, are denoted
$$\frac{\partial^2 f}{\partial \mu^2}(y;\mu), \quad \frac{\partial^2 f}{\partial y \partial \mu}(y;\mu)$$
etc.

Definition 1.1 *Let $\Omega \subseteq C \subseteq \mathbf{R}$ be intervals with Ω open. A function $d: C \times \Omega \to \mathbf{R}$ is called a **unit deviance** if it satisfies*

$$d(y;y) = 0 \quad \forall y \in \Omega \tag{1.1}$$

and

$$d(y;\mu) > 0 \quad \forall y \neq \mu. \tag{1.2}$$

*A unit deviance d is called **regular** if $d(y;\mu)$ is twice continuously differentiable with respect to (y,μ) on $\Omega \times \Omega$ and satisfies*

$$\frac{\partial^2 d}{\partial \mu^2}(\mu;\mu) > 0 \quad \forall \mu \in \Omega. \tag{1.3}$$

*The **unit variance function** $V: \Omega \to \mathbf{R}_+$ of a regular unit deviance is defined by*

$$V(\mu) = \frac{2}{\dfrac{\partial^2 d}{\partial \mu^2}(\mu;\mu)}.$$

Unit deviances may often be constructed on the basis of log likelihoods. A log likelihood $l(y;\mu)$ such that the maximum likelihood estimate of μ is y leads to the unit deviance

$$d(y;\mu) = c\{l(y;y) - l(y;\mu)\}, \tag{1.4}$$

DEFINITIONS

for a given constant c, provided that (1.2) is satisfied. Such log likelihoods are examples of **yokes** (Blæsild, 1987, 1991; Barndorff-Nielsen, 1989). As shown in Exercise 1.2, yokes and unit deviances are closely related.

The normal unit deviance $d(y; \mu) = (y - \mu)^2$ is obviously regular with unit variance function $V(\mu) = 1$. An example of a unit deviance that is not regular is

$$d(y; \mu) = |y - \mu|. \quad (1.5)$$

Not that the definition does not require the unit deviance to be symmetric in its two arguments, although it is locally symmetric for y near μ, see equation (1.22) below.

We now introduce reproductive dispersion models.

Definition 1.2 *A **reproductive dispersion model** $\mathrm{DM}(\mu, \sigma^2)$ with **position parameter** μ and **dispersion parameter** σ^2 is a family of distributions whose probability density functions with respect to a suitable measure may be written in the form*

$$p(y; \mu, \sigma^2) = a(y; \sigma^2) \exp\left\{-\frac{1}{2\sigma^2} d(y; \mu)\right\}, \quad y \in C, \quad (1.6)$$

*where $a \geq 0$ is a suitable function, d is a unit deviance on $C \times \Omega$, $\mu \in \Omega$ and $\sigma^2 > 0$. A dispersion model density of the form (1.6) is said to be expressed in **standard form**.*

Here it is understood, because d is a unit deviance on $C \times \Omega$, that $\Omega \subseteq C$, and that Ω is an interval. Also, note that the support of $\mathrm{DM}(\mu, \sigma^2)$ depends on σ^2 only. We require (1.6) to hold throughout the convex support C, which implies that $a(y; \sigma^2)$ is zero outside the support of $\mathrm{DM}(\mu, \sigma^2)$. Note that, for the unit deviance from (1.6), the log likelihood for μ when σ^2 is known is $l(y; \mu) = -d(y; \mu)/(2\sigma^2)$, and this log likelihood is precisely of the form required for the construction (1.4).

Further definitions

It is useful at this point to anticipate some definitions from later chapters. These definitions will be given in preliminary form here, and more precise definitions will be given in the relevant chapters.

We call (1.6) a **regular proper dispersion model**, denoted $\mathrm{PD}(\mu, \sigma^2)$, if the unit deviance d is regular, $S = C = \Omega$, and (1.6) takes the form

$$p(y; \mu, \sigma^2) = a(\sigma^2) V^{-\frac{1}{2}}(y) \exp\left\{-\frac{1}{2\sigma^2} d(y; \mu)\right\}, \quad (1.7)$$

for $y, \mu \in \Omega$, for a suitable function a, where V is the unit variance function. The qualifier 'regular' refers to the regularity of the unit deviance. In the remainder of the present chapter, this qualifier is dropped for simplicity, and we refer to (1.7) as simply a *proper dispersion model*. Proper dispersion models will be studied further in Chapter 5.

We call (1.6) a **reproductive exponential dispersion model**, denoted $\mathrm{ED}(\mu, \sigma^2)$, if the unit deviance takes the form

$$d(y; \mu) = yf(\mu) + g(\mu) + h(y), \tag{1.8}$$

for suitable functions f, g and h. A **natural exponential family**, denoted $\mathrm{NE}(\mu)$, is a family with densities of the form

$$c(y) \exp\left\{-\frac{1}{2} d(y; \mu)\right\},$$

where d is of the form (1.8). In particular, a reproductive exponential dispersion model with σ^2 known is a natural exponential family. We may rewrite the density of a natural exponential family in the more familiar form

$$c_1(y) \exp\left\{yf_1(\mu) + g_1(\mu)\right\}, \tag{1.9}$$

for suitable c_1, f_1 and g_1. Natural exponential families and exponential dispersion models will be studied in Chapters 2 and 3 respectively.

The normal distribution is obviously a proper dispersion model. Moreover, the expansion

$$(y - \mu)^2 = -2y\mu + \mu^2 + y^2$$

shows that it is also a reproductive exponential dispersion model.

Consider a reproductive dispersion model $Y \sim \mathrm{DM}\left(\mu, \sigma^2\right)$. The transformation from Y to Z defined by

$$Z = \frac{Y}{\sigma^2}$$

is called the **duality transformation**, and plays an important role for exponential dispersion models. Note that the duality transformation depends on the value of the dispersion parameter σ^2. The density of the variable Z is

$$p^*(z; \xi, \sigma^2) = a^*(z; \sigma^2) \exp\left\{-\frac{1}{2\sigma^2} d(z\sigma^2; \xi\sigma^2)\right\}, \tag{1.10}$$

DEFINITIONS

Table 1.1. *Summary of dispersion models*

	Exponential	Proper
Reproductive	$\mathrm{ED}(\mu,\sigma^2)$ (Ch. 3)	$\mathrm{PD}(\mu,\sigma^2)$ (Ch. 5)
Additive	$\mathrm{ED}^*(\theta,\lambda)$ (Ch. 3)	—
σ^2 known	$\mathrm{NE}(\mu)$ (Ch. 2)	—

where $\xi = \mu/\sigma^2$ and

$$a^*(z;\sigma^2) = \begin{cases} \sigma^2 a(z\sigma^2;\sigma^2) & \text{continuous case} \\ a(z\sigma^2;\sigma^2) & \text{discrete case.} \end{cases}$$

We call (1.10) an **additive dispersion model**; in particular we call (1.10) an **additive exponential dispersion model** if $Y \sim \mathrm{ED}(\mu,\sigma^2)$. In the latter case we use the symbol $Z \sim \mathrm{ED}^*(\theta,\lambda)$, where $\theta = f\left(\xi\sigma^2\right) = f(\mu)$ from (1.8) and $\lambda = 1/\sigma^2$.

By a **transformational dispersion model**, we understand a dispersion model where μ is the group parameter for a group of transformations leaving σ^2 is invariant. In Section 1.4.2 we discuss the special case where μ is either a location or scale parameter. This type of model is particularly important in connection with (regular) proper dispersion models, see Section 5.2.2. Dispersion models not of the transformational form are called **non-transformational dispersion models**.

The statistical inference for reproductive dispersion models is in some sense more appealing than for additive dispersion models. However, additive exponential dispersion models turn out to be important for discrete data because many important families of discrete distributions have this form. Additive proper dispersion models will not be considered further here.

Table 1.1 summarizes the main types of dispersion models, with references to the appropriate chapters.

Note on terminology

When no confusion can arise, we refer to reproductive dispersion models as simply dispersion models in the following. Most properties of reproductive dispersion models may be translated into the additive case by the duality transformation.

1.2.2 Comments on definitions

We now present some general interpretations of the above definitions, as well as some explanations of the terminology used.

World map of dispersion models

Exponential and proper dispersion models are the two main classes of dispersion models. It turns out that examples of dispersion models outside these two classes are rare, and the overlap between the two classes is small. In fact, by Theorem 5.6 in Chapter 5, the overlap consists of exactly three models, namely the normal, gamma and inverse Gaussian distributions (the latter is introduced in Section 4.2.3). The common structure of dispersion models is mainly important for their statistical properties, as we shall see in Section 1.3. However, from the point of view of distributional properties, the class of dispersion models serves mainly as a convenient umbrella class for exponential and proper dispersion models. While each of these two subclasses enjoys many important properties, few such properties are shared by both classes. Exceptions to this rule are properties of the probability density functions derived from those of the unit deviance, as explained below.

Figure 1.2 illustrates the different types of dispersion models, in the form of a 'world map'. Proper and exponential dispersion models are represented in the figure by the two hemispheres, and transformational and non-transformational models are separated by the equator. As already mentioned, dispersion models that are neither proper nor exponential are rare, and are hence, appropriately, illustrated in the figure as having fallen off the edge of the world! Among the three models in the intersection between proper and exponential dispersion models, the normal and gamma distributions are transformational, whereas the inverse Gaussian is not, see Section 5.2.3.

At this point it should be made clear that dispersion models outside of the classes of exponential and proper dispersion models are still not well understood, mainly for lack of examples of this kind. We shall later present methods for constructing exponential and proper dispersion models, respectively, but no such method currently exists for producing dispersion models outside of these two classes.

DEFINITIONS

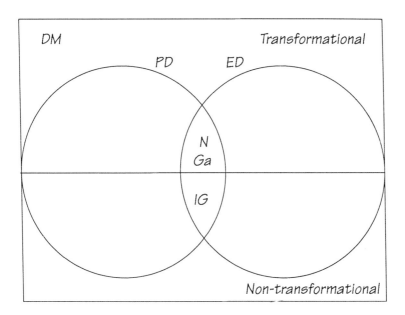

Figure 1.2. *World map of dispersion models* (DM). *Exponential dispersion models* (ED) *and proper dispersion models* (PD) *are indicated by the two hemispheres. Transformational models and non-transformational models are indicated by the northern and southern hemispheres, respectively.*

Interpretation of density

We now consider the shape of the dispersion model density

$$a(y;\sigma^2)\exp\left\{-\frac{1}{2\sigma^2}d(y;\mu)\right\}. \tag{1.11}$$

The general idea of the definition is that, with d being a measure of squared distance, the second factor of (1.11) tends to give a mode point of the density near μ. The smaller the value of σ is, the higher and more narrow the peak of this mode will be. This makes μ and σ somewhat analogous to location and scale, respectively. In particular, the parameter σ^2 quantifies the notion of *dispersion* mentioned by Fisher (1953) in the opening quote of this chapter.

The above interpretation of μ and σ depends on the detailed properties of $d(y;\mu)$ as a squared distance. In many cases, $d(y;\mu)$ is in fact monotone as a function of y on each side of μ, but no such assumption is necessary in the following, an exception being the definition of residuals in Section 3.6.

The interpretation of μ and σ is somewhat blurred by the fact that $a(y; \sigma^2)$ may depend on y. For this reason, we use the more general terminology *position* and *dispersion* instead of *location* and *scale*, respectively. We might have stressed this analogy more by using the term *position-dispersion models*, but we maintain the simpler *dispersion models*.

A crucial aspect of the definition is that (1.11) must be a probability density function for *all* values of the parameters μ and σ^2. This means that only very special functions a and d can appear in the definition. While the definition of a unit deviance in Definition 1.1 is very general, the class of unit deviances that can appear in connection with dispersion models is much more restricted. Specifically, $-d(y; \cdot)/(2\sigma^2)$ is then a log likelihood for each fixed value of σ^2. The general definition of unit deviance is, however, useful for other purposes in the following.

Another crucial aspect of the definition is that $a(y; \sigma^2)$ does not depend on μ. This generalizes the second property of the normal distribution mentioned at the beginning of Section 1.2, although in contrast to the normal case, $a(y; \sigma^2)$ may depend on y.

We note that, in order to show that a given two-parameter family of distributions is a dispersion model, we must find a parametrization (μ, σ^2) that brings the density into the form (1.11). The requirements that $\Omega \subseteq C$ and $d(y; y) = 0$ essentially determine the parameter μ uniquely, because this condition implies that the maximum likelihood estimate of μ is y (see Exercise 1.3). The parameter σ^2 is determined only up to a multiplicative factor, because multiplying both d and σ^2 by the same constant leaves the dispersion model as such unchanged. In most examples, σ^2 varies in $\mathbf{R}_+$, but other domains are possible, for example a finite interval $(0, \sigma_0^2)$.

Justification of terminology

The word 'unit' in the terms 'unit deviance' and 'unit variance function' is used here both in the statistical sense of 'observational units' (see equation (1.12) below) and in the sense of corresponding to standardized forms of the functions $\sigma^{-2}d(y; \mu)$ and $\sigma^2 V(\mu)$, respectively, with $\sigma^2 = 1$.

The terminology 'variance function' refers to the role of $\sigma^2 V(\mu)$ as the asymptotic variance of Y, see Theorem 1.5 below. For exponential dispersion models, we show in Chapter 3 that the exact

DEFINITIONS

variance of Y is $\sigma^2 V(\mu)$ for any value of σ^2. In this case, the variance function hence summarizes how the variance behaves as a function of μ. Like the unit deviance, the unit variance function plays a crucial role in the following. Being essentially the curvature of the unit deviance at its minimum, the unit variance function summarizes the main feature of the shape of the unit deviance. Furthermore, the shape of the unit variance function itself provides a useful summary of the degree and type of non-normality of the corresponding dispersion model.

The terminology 'reproductive' for reproductive exponential dispersion models refers to the property that the distribution of the sample mean $\bar{Y}$ for a random sample of size n from the model belongs to the model itself, see formula (3.16) in Chapter 3. For an arbitrary reproductive dispersion model, the asymptotic variance for the sample mean $\bar{Y}$ is $(\sigma^2/n)V(\mu)$ for σ^2 small, corresponding to replacing σ^2 by σ^2/n in the asymptotic variance. This is a kind of asymptotic reproductive property of the model, supporting the terminology 'reproductive dispersion model'.

Similarly, the terminology 'additive' for additive exponential dispersion models refers to the property that the distribution of the sample sum Z_+ for a random sample of size n from the family belongs to the family itself. This follows from the above reproductive property by the duality transformation. An arbitrary additive dispersion model has asymptotic variance $\sigma^{-2}V(\sigma^2\xi)$ for σ^2 small, where $\xi = \mu/\sigma^2$ is the asymptotic mean. Hence, the asymptotic mean of Z_+ is $\xi' = n\xi$ and the asymptotic variance is

$$\frac{n}{\sigma^2}V\left(\frac{\sigma^2}{n}\xi'\right),$$

corresponding to replacing σ^2 by σ^2/n in the asymptotic variance. We call this the asymptotic additive property of the model, supporting the terminology 'additive dispersion model'.

Small and large dispersions

We show in Theorem 1.5 below that, under regularity conditions, dispersion models are approximately normal for small values of the dispersion parameter σ^2. Hence, data with small dispersions may be analysed by means of the normal distribution, using the traditional method of transformations, where a transformation of the data is sought that will linearize the model and stabilize the variance. However, as Fisher (1953) wrote in the above quote: 'It is

... of some little mathematical interest to consider how the theory would have had to be developed if the observations under discussion had in fact involved errors so large that the actual topology had had to be taken into account.' He goes on to say that '... there are in nature vectors with such large natural dispersions.' Encouraged by this, we proceed to develop the theory of dispersion models in order to take such 'large dispersions' into account.

1.3 Analysis of deviance for generalized linear models

We now digress briefly to discuss the main ideas of analysis of deviance for generalized linear models with error distributions from the class of dispersion models. This topic provides important motivation for our study of dispersion models, but will otherwise not be pursued further in the present book. The interested reader is referred to other sources for details, such as for example Jørgensen (1987a, 1997) and McCullagh and Nelder (1989).

Consider independent observations $Y_1, \ldots, Y_n$, with distributions $Y_i \sim \mathrm{DM}(\mu_i, \sigma^2)$, where $\mathrm{DM}(\mu, \sigma^2)$ denotes a dispersion model, so that σ^2 is unknown, but common for all observations. A **generalized linear model** is defined by further assuming that μ_i satisfies

$$g(\mu_i) = \mathbf{x}_i^\top \boldsymbol{\beta}$$

for $i = 1, \ldots, n$. Here g is a known link function, $\mathbf{x}_i$ is a vector of covariates, and $\boldsymbol{\beta}$ a vector of unknown regression parameters.

We define the (total) **deviance** for the parameter vector $\boldsymbol{\mu} = (\mu_1, \ldots, \mu_n)^\top$ based on the sample $\mathbf{y} = (y_1, \ldots, y_n)^\top$ by

$$D(\mathbf{y}; \boldsymbol{\mu}) = \sum_{i=1}^{n} d(y_i; \mu_i). \tag{1.12}$$

Given the interpretation of the unit deviance $d(y; \mu)$ as analogous to the squared distance $(y - \mu)^2$, the total deviance may be considered a generalization of the familiar residual sum of squares from normal theory. It is not difficult to see that maximum likelihood estimation for the parameter $\boldsymbol{\beta}$ of a generalized linear model corresponds to minimizing the total deviance, thus generalizing least-squares estimation.

More generally, we may consider a smooth subset $\Omega_1 \subseteq \Omega^n$, where Ω^n is the full parameter space for $\boldsymbol{\mu}$. We then define the deviance for the hypothesis $H_1 : \boldsymbol{\mu} \in \Omega_1$ by

$$D_1 = \inf_{\boldsymbol{\mu} \in \Omega_1} D(\mathbf{y}; \boldsymbol{\mu}).$$

Let D_2 denote the deviance for a sub-hypothesis H_2 of H_1, corresponding to the smooth subset $\Omega_2 \subseteq \Omega_1$.

In cases where σ^2 is known, in particular for natural exponential families, inference may be based on the difference between the two deviances, $D_2 - D_1$. This is simply the log likelihood ratio statistic, which has an asymptotic χ^2 distribution, generalizing ideas from logistic regression and analysis of loglinear models for contingency tables.

When σ^2 is unknown, inference may be based on the following F-statistic:

$$F = \frac{(D_2 - D_1)/(f_2 - f_1)}{D_1/f_1}, \tag{1.13}$$

where f_1 and f_2 denote the degrees of freedom of Ω_1 and Ω_2, respectively. In the case of normal-theory linear models, where D_1 and D_2 are the residual sums of squares corresponding to Ω_1 and Ω_2, respectively, F is the usual F-statistic from analysis of variance and regression. In this case, F follows an $F(f_2 - f_1, f_1)$ distribution under H_2.

For arbitrary dispersion models, we may show, using the asymptotic normality of the distribution for σ^2 small, that the F-statistic (1.13) is asymptotically distributed as $F(f_2 - f_1, f_1)$ when σ^2 is small. This serves as a basis for testing H_2 under H_1. Other versions of the F-statistic exist such that the F-approximation applies for large samples, as well as for small values of σ^2. One may generalize other familiar procedures, such as the t-test, in a similar way.

The above inference methods based on the deviance make up the essential ingredients of the analysis of deviance. The results are meant to give an impression of how the analogies between the normal distribution and dispersion models extend to analogies between analysis of variance and analysis of deviance. Similar ideas apply, to some extent, for inference in additive dispersion models.

1.4 Examples

We now consider a number of important examples of dispersion models that illustrate the different distribution types from the list on p. 2. We accompany this with some historical remarks on the development of generalized linear models and dispersion models. Most of the examples will be considered in more detail in subsequent chapters.

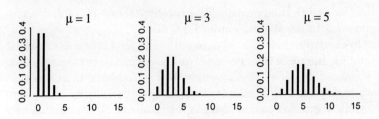

Figure 1.3. *Some Poisson probability functions.*

1.4.1 Discrete data

Historically, discrete data analysis was developed relatively early on compared with other areas of non-normal data analysis. One reason is the frequency with which discrete data are met in practice. Another reason may be that the two main examples of discrete distributions, the Poisson and the binomial, are both natural exponential families, whose statistical analysis is relatively straightforward.

Poisson distribution

The Poisson distribution Po(μ) with mean μ has probability function
$$p(y;\mu) = \frac{\mu^y}{y!}e^{-\mu} = \frac{1}{y!}\exp\left(y\log\mu - \mu\right)$$
for $y \in \mathbf{N}_0$. The last formula shows that it is a natural exponential family (compare with equation (1.9)). Figure 1.3 shows some examples of Poisson probability functions, illustrating the nature of μ as a position parameter. The unit deviance of the Poisson distribution is
$$d(y;\mu) = 2\left(y\log\frac{y}{\mu} - y + \mu\right),$$
and the corresponding unit variance function is $V(\mu) = \mu$.

The Poisson distribution is in a sense the 'normal' distribution for count data, because it is the limiting distribution for counts of 'rare events', in parallel with the Central Limit Theorem, where the normal distribution appears as the limiting distribution for sums of many small contributions. The arguments leading to the Poisson limit may be summarized as follows (see Feller, 1968, pp. 153–156). If members of a very large population are each exposed to a certain type of rare event, independently of each other, and each with the same very small probability, then the number of events in a given time interval follows the Poisson distribution.

However, in contrast to the normal distribution, the Poisson distribution has but a single parameter, having no dispersion parameter. This makes it less flexible than the normal for fitting actual data, and raises the question of whether an appropriate two-parameter distribution such as the negative binomial should be used routinely for analysing count data.

Binomial distribution

The other main type of discrete data is the case of binomial counts, corresponding to the support $S = \{0, \ldots, m\}$, where we record the number, Y, of events of a certain type occurring in m independent trials. If the event under consideration has constant probability μ, then Y follows the binomial distribution $\text{Bi}(m, \mu)$, which has probability function of the form

$$p(y; \mu, m) = \binom{m}{y} \mu^y (1-\mu)^{m-y},$$

for $y = 0, \ldots, m$. Some examples of binomial probability functions are shown in Figure 1.4. This plot illustrates the role of μ as a parameter of position, for each fixed m.

By writing the binomial probability function as follows:

$$p(y; \mu, m) = \binom{m}{y} \exp\left\{ y \log \frac{\mu}{1-\mu} + m \log(1-\mu) \right\},$$

we find that it is a natural exponential family for each fixed value of m. However, since m is known in most practical situations, the binomial distribution is essentially a one-parameter distribution, sometimes limiting its usefulness for fitting actual data. Again, this leads to the question of finding appropriate two-parameter alternatives to the binomial distribution, such as, for example, binomial mixtures.

Further comments on discrete case

The binomial and Poisson were the two main examples of discrete error distributions for generalized linear models considered by Nelder and Wedderburn (1972). Around the same time, several books on loglinear and logistic models appeared, such as Cox (1970), Haberman (1974) and Bishop, Fienberg and Holland (1975).

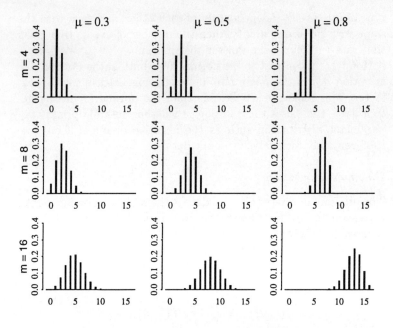

Figure 1.4. *Some binomial probability functions.*

This, together with the release of the computer program GLIM for generalized linear models soon after, helps to explain why Nelder and Wedderburn's ideas caught on relatively quickly. See Agresti (1990) for further historical notes on discrete data analysis.

The Poisson distribution turns out to be an additive exponential dispersion model in a certain formal sense (see Section 3.3.3). In practice, however, both the Poisson and the binomial lack the dispersion parameter that makes other dispersion models such as the normal so flexible for fitting data. As we shall see in Chapter 3, the only possible discrete exponential dispersion models with a genuine dispersion parameter are additive models, such as for example the negative binomial distribution (Section 3.3.5) and a new class of Tweedie-Poisson mixtures to be introduced in Section 4.6. However, inferential techniques are not yet as well developed for additive dispersion models as they are for reproductive dispersion models. Hence, fully flexible methods for analysis of discrete data are still not readily available.

1.4.2 Location-dispersion models

Many early attempts at dealing with continuous non-normal data have involved location-scale models, for example the least-squares method. Sweeting (1984) showed that, when suitably parametrized, location-scale regression models have much in common with generalized linear models. We now consider dispersion models that involve a location or a scale parameter, and consider the relationship with location-scale models.

Let $d : \mathbf{R} \mapsto \mathbf{R}$ be a non-negative function satisfying $d(0) = 0$ and $d(y) > 0$ for $y \neq 0$. Then $d(y - \mu)$ is a unit deviance. If the integral

$$\frac{1}{a(\sigma^2)} = \int_{-\infty}^{\infty} \exp\left\{-\frac{1}{2\sigma^2} d(y)\right\} dy$$

is finite for σ^2 in some interval $(0, \sigma_0^2)$, we may hence define a dispersion model on $\mathbf{R}$ by

$$p(y; \mu, \sigma^2) = a(\sigma^2) \exp\left\{-\frac{1}{2\sigma^2} d(y - \mu)\right\}. \quad (1.14)$$

The parameter μ is a location parameter, and we call the model (1.14) a **location-dispersion model**. If d is twice continuously differentiable and $d''(0) > 0$, the unit deviance $d(y - \mu)$ is regular, and the corresponding unit variance function is constant. In this case (1.14) is a proper dispersion model. Note that we may rescale the unit deviance so that the unit variance function becomes $V(\mu) = 1$, the unit variance function of the normal distribution, showing that many different dispersion models may share the same unit variance function. However, a natural exponential family is characterized by its unit variance function (see Theorem 2.11), and similarly for exponential dispersion models. Hence, the only exponential dispersion model with constant variance function on $\mathbf{R}$ is the normal distribution.

In Exercise 1.9 we consider the functions $d(y) = |y|^\rho$. In particular, the cases $\rho = 2$ and $\rho = 1$ give the normal and Laplace (double exponential) distributions, respectively.

The analogy between location-dispersion models and location-scale models is easily brought out if we write a location-scale model in the form

$$p(y; \mu, \sigma^2) = \sigma^{-1} \exp\left\{-d\left(\frac{y-\mu}{\sigma}\right)\right\}, \quad (1.15)$$

where $-d$ is the logarithm of a given density function. The crucial difference between the two is that in (1.15), σ^{-1} appears in the argument of d, whereas in (1.14) σ^{-2} is a multiplier of d.

If we apply the transformation $z = \exp y$ together with the reparametrization $\xi = \exp \mu$ to the location-dispersion model (1.14), we obtain a dispersion model of the form

$$p(z;\xi,\sigma^2) = a(\sigma^2)z^{-1}\exp\left\{-\frac{1}{2\sigma^2}d_0\left(\frac{z}{\xi}\right)\right\}$$

for $z > 0$, where $d_0(z) = d(\log z)$. We call this model a **scale-dispersion model**. If d is regular, this is a proper dispersion model with unit variance function proportional to μ^2. Note again that there are many proper dispersion models that have the same unit variance function.

Location- and scale-dispersion models are special cases of dispersion models generated by groups of transformations. Such models will be considered in Section 5.2.

1.4.3 Positive data

An important distribution for positive data is the gamma distribution with density function for $y > 0$ given by

$$\frac{\psi^\lambda}{\Gamma(\lambda)}y^{\lambda-1}e^{-\psi y}, \tag{1.16}$$

where the parameters λ and ψ are positive. Indeed, the gamma was the only continuous distribution mentioned by Nelder and Wedderburn (1972) other than the normal. In earlier times, the transformation method was the main comprehensive technique available for analysis positive data, a key reference being Box and Cox (1964).

Defining $\mu = \lambda/\psi$ (the mean) and $\sigma^2 = 1/\lambda$ (the squared coefficient of variation), the density may be written as follows:

$$p(y;\mu,\sigma^2) = a(y;\sigma^2)\exp\left\{-\frac{1}{2\sigma^2}2\left(\frac{y}{\mu} - \log\frac{y}{\mu} - 1\right)\right\},$$

where a is defined by

$$a(y;1/\lambda) = \frac{\lambda^\lambda e^{-\lambda}}{\Gamma(\lambda)}y^{-1}. \tag{1.17}$$

This shows that the gamma distribution is a dispersion model with position μ, dispersion σ^2 and unit variance function $V(\mu) = \mu^2$. It is in fact a proper dispersion model and a scale-dispersion model, since (1.17) factorizes into a function of σ^2 times $V^{-1/2}(y) = y^{-1}$.

EXAMPLES

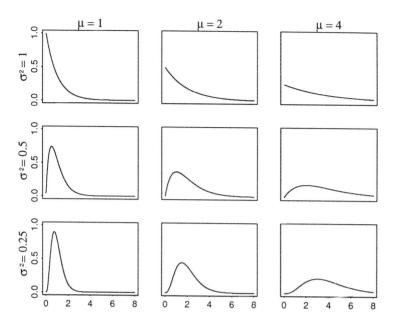

Figure 1.5. *Some gamma densities.*

The unit deviance is

$$d(y;\mu) = 2\left(\frac{y}{\mu} - \log\frac{y}{\mu} - 1\right),$$

which shows that the gamma distribution is also an exponential dispersion model. The gamma distribution is denoted by the symbol $\text{Ga}(\mu,\sigma^2)$.

Figure 1.5 shows some examples of gamma density functions. This (and similar plots in the following) are arranged in three columns, each corresponding to a given value of μ, and three rows, each corresponding to a given value of σ^2. This arrangement is meant to bring out the interpretation of μ and σ^2 as position and dispersion parameters, respectively, and to show the variety of shapes that the family of densities can attain.

The behaviour of the densities in Figure 1.5 is fairly typical for many dispersion models. For small values of σ^2, the density is clearly unimodal with mode point near μ. The three plots with $\sigma^2 = 1$, which are all exponential distributions, illustrate the case where the dispersion is so large that the interpretation of μ as 'position' becomes blurred, essentially because the large value of the dispersion parameter squeezes a lot of probability mass down towards the origin. What we mean by a 'large' value of the dispersion parameter here is thus a value for which the non-normality of the distribution becomes evident. Further details about the gamma distribution are given in Section 3.3.2.

1.4.4 Positive data with zeros

Positive data with zeros provides an example of data for which there was initially no solution available within the framework of generalized linear models, and where the transformation method generally does not work well. This case corresponds to the support $S = \mathbf{R}_0$, where the outcome zero occurs with positive probability, while the distribution is continuous for positive outcomes. An example of such data are measurements of precipitation, where measured amounts for wet periods show continuous variation, while dry periods are recorded as a zero. Another example is the total claim on an insurance policy over a fixed time interval, where the total claim may be either positive if claims were made in the period, or zero if no claims were made. A third example corresponds to a measuring device that records a zero whenever the quantity measured falls below a certain threshold.

The systematic study of exponential dispersion models initiated by Jørgensen (1987a) opened up the possibility for analysing a wide range of data beyond what had been suggested by Nelder and Wedderburn (1972). The prime examples of this are the compound Poisson exponential dispersion models, which are suitable for positive data with zeros. The unit deviances for these models are given by

$$d_p(y;\mu) = 2\left\{\frac{y^{2-p}}{(1-p)(2-p)} - \frac{y\mu^{1-p}}{1-p} + \frac{\mu^{2-p}}{2-p}\right\},$$

where the parameter p belongs to the interval $(1, 2)$. The corresponding unit variance function is $V(\mu) = \mu^p$. For technical reasons, we defer the discussion of these models to Chapter 4.

1.4.5 Directions

One of the shortcomings of exponential dispersion models is that there are no adequate exponential dispersion models available for data with bounded support, such as angles or proportions. The technical explanation for this will be given in Chapter 3. In outline, the argument is that we require the parameter σ^2 to vary continuously, which in turn implies that the distributions in the family are infinitely divisible, and such distributions have unbounded support.

The first step towards accommodating such data within the generalized linear models framework was taken by Jørgensen (1983, 1984), who introduced what we now call dispersion models, and showed how to extend the estimation and inference techniques for generalized linear models to this wider class of models.

We now consider the case of **directional data**, meaning data in the interval $[0, 2\pi)$, representing angular measurements or geographical directions. We may think of the interval $[0, 2\pi)$ as representing either the unit circle or $\mathbf{R} \bmod 2\pi$.

One of the distributions included in the framework of dispersion models is the **von Mises distribution** for directional data. The von Mises distribution is defined by the density function

$$\frac{1}{2\pi I_0(\lambda)} \exp\{\lambda \cos(y - \mu)\}, \tag{1.18}$$

for $0 \le y < 2\pi$, where $\mu \in [0, 2\pi)$, $\lambda > 0$ and I_0 denotes the modified Bessel function given by

$$I_0(\lambda) = \frac{1}{2\pi} \int_0^{2\pi} \exp(\lambda \cos y) \, dy.$$

We shall meet other Bessel functions later on; a good reference for their properties is Abramowitz and Stegun (1972, Chapters 9, 10).

The von Mises distributions (1.18) are a proper dispersion model, and

$$d(y; \mu) = 2\{1 - \cos(y - \mu)\}$$

is a unit deviance on $[0, 2\pi) \times (0, 2\pi)$. The von Mises density function may be expressed in standard form as follows:

$$p(y; \mu, \sigma^2) = \frac{e^{\sigma^{-2}}}{2\pi I_0(\sigma^{-2})} \exp\left[-\frac{1}{2\sigma^2} 2\{1 - \cos(y - \mu)\}\right],$$

with position μ and dispersion $\sigma^2 = 1/\lambda$, denoted by the symbol vM(μ, σ^2). We note that μ is the mode point of the density. The unit variance function is $V(\mu) = 1$, the same as for the normal distribution, except for the domain, which is $(0, 2\pi)$. Figure 1.6 shows some plots of von Mises density functions.

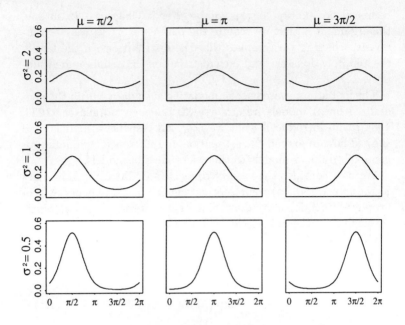

Figure 1.6. *Some von Mises densities.*

1.4.6 Proportions

We now turn to distributions suitable for **proportions**, meaning data on the unit interval, also known as **compositional data** or **continuous proportions**. Superficially, there is little difference between data on the unit interval and the interval $[0, 2\pi)$, but while directions generally have no canonical origin, the endpoints 0 and 1 have a clearly defined significance for proportions.

One of the simplest distributions for proportions is the beta distribution, but this distribution is apparently not a dispersion model (see Exercise 1.4). The initial studies of dispersion models did not reveal any examples suitable for proportions, and provided little in the way of tools for constructing such examples. However, such tools became available with the introduction of proper dispersion models, see Jørgensen (1997), and led to the discovery of the **simplex distribution**, introduced by Barndorff-Nielsen and Jørgensen (1991).

The univariate simplex distribution $S^-(\mu, \sigma^2)$ with parameters $\mu \in (0, 1)$ and $\sigma^2 > 0$ is defined by the probability density function

EXAMPLES

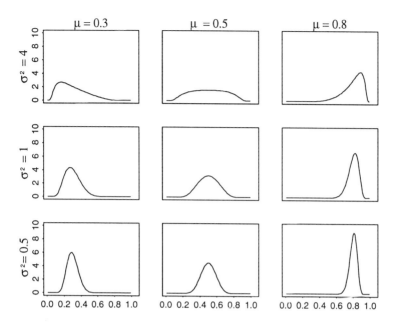

Figure 1.7. *Some simplex densities.*

$$p(y; \mu, \sigma^2) = \left[2\pi\sigma^2\{y(1-y)\}^3\right]^{-\frac{1}{2}} \exp\left\{-\frac{1}{2\sigma^2}d(y;\mu)\right\}, \quad (1.19)$$

for $0 < y < 1$, where

$$d(y;\mu) = \frac{(y-\mu)^2}{y(1-y)\mu^2(1-\mu)^2} \quad (1.20)$$

is a unit deviance. The corresponding unit variance function is $V(\mu) = \mu^3(1-\mu)^3$ for $\mu \in (0,1)$. The form of the density (1.19) shows that the simplex distribution is a proper dispersion model.

The unit deviance (1.20) resembles a squared distance because of the numerator $(y-\mu)^2$, but d is not a quadratic form in μ and y because the denominator of (1.20) involves both μ and y. Figure 1.7 shows some examples of simplex density functions. These plots show that the simplex distribution is a case where the parameters μ and σ^2 have very clear interpretations as position and dispersion parameters, respectively, in spite of the fact that the density is confined to a bounded interval. A more general class of simplex distributions is studied in Section 5.3.3.

1.5 Properties of dispersion models

We now consider some general properties of unit deviances and dispersion models. In particular, we will consider the saddlepoint approximation in Section 1.5.3.

1.5.1 Unit deviance

In spite of the simplicity of its definition, a regular unit deviance has some useful properties regarding its behaviour near its minimum.

Lemma 1.1 *A regular unit deviance satisfies*
$$\frac{\partial^2 d}{\partial y^2}(\mu;\mu) = \frac{\partial^2 d}{\partial \mu^2}(\mu;\mu) = -\frac{\partial^2 d}{\partial \mu \partial y}(\mu;\mu) \quad \forall \mu \in \Omega.$$

Proof. By definition, a regular unit deviance satisfies the conditions
$$d(y;y) = d(\mu;\mu) = 0 \quad \text{and} \quad d(y;\mu) > 0 \quad \text{for} \quad y \neq \mu.$$
Hence, the function $d(y;\,\cdot\,)$ has a unique minimum at y, and similarly $d(\,\cdot\,;\mu)$ has a unique minimum at μ, implying that for all $\mu \in \Omega$
$$\frac{\partial d}{\partial \mu}(\mu;\mu) = 0 \quad \text{and} \quad \frac{\partial d}{\partial y}(\mu;\mu) = 0 \ .$$
By differentiating the first of the above equations with respect to μ we obtain, by the chain rule,
$$\frac{\partial^2 d}{\partial \mu^2}(\mu;\mu) + \frac{\partial^2 d}{\partial \mu \partial y}(\mu;\mu) = 0,$$
and similarly from the second equation,
$$\frac{\partial^2 d}{\partial y^2}(\mu;\mu) + \frac{\partial^2 d}{\partial \mu \partial y}(\mu;\mu) = 0.$$
The result now follows by combining these two equations. $\square$

The lemma gives us three equivalent expressions for the unit variance function,
$$V(\mu) = \frac{2}{\dfrac{\partial^2 d}{\partial \mu^2}(\mu;\mu)} = \frac{2}{\dfrac{\partial^2 d}{\partial y^2}(\mu;\mu)} = -\frac{2}{\dfrac{\partial^2 d}{\partial \mu \partial y}(\mu;\mu)}. \tag{1.21}$$

PROPERTIES OF DISPERSION MODELS 25

Furthermore, the lemma and (1.21) imply the following second-order expansion of d near its minimum:

$$d(\mu_0 + x\delta; \mu_0 + m\delta) = \frac{\delta^2}{V(\mu_0)}(x-m)^2 + o(\delta^2). \qquad (1.22)$$

This expansion shows that a regular unit deviance behaves approximately as the normal unit deviance near its minimum, with curvature given by the reciprocal of the unit variance function.

1.5.2 Transformations

We now consider transformations of dispersion models; in particular we introduce the notion of a variance-stabilizing transformation.

Consider a unit deviance d on $C \times \Omega$, and a one-to-one transformation $f : C \to C_0$. A new unit deviance d_0 may then be defined on $f(C) \times f(\Omega)$ by

$$d_0(z; \xi) = d\left\{f^{-1}(z); f^{-1}(\xi)\right\}.$$

In other words, we obtain a new unit deviance by transforming both y and μ by the same transformation, letting $z = f(y)$ and $\xi = f(\mu)$. We refer to d_0 as the unit deviance obtained from d by **reparametrization** by f.

If Y follows a dispersion model with unit deviance d, consider the transformation from Y to $Z = f(Y)$ with f being monotone and differentiable. Then Z follows a dispersion model with unit deviance d_0, as seen from the following expression for the density function of Z in the continuous case:

$$p_Z(z; \xi, \sigma^2) = \frac{a\left\{f^{-1}(z); \sigma^2\right\}}{|f'\{f^{-1}(z)\}|} \exp\left\{-\frac{1}{2\sigma^2} d_0(z; \xi)\right\}. \qquad (1.23)$$

A similar formula applies in the discrete case, and corresponds to ignoring the Jacobian in (1.23). In order to preserve the dispersion model structure, we must hence simultaneously transform Y and reparametrize μ by the same transformation.

In this sense the dispersion model structure is preserved under arbitrary monotone differentiable transformations. This may be compared with the case of location-scale models, where the location-scale structure is preserved in a similar way under arbitrary linear transformations.

Now assume that $f : C \to C_0$ is twice continuously differentiable and satisfies $|f'(y)| > 0$ on Ω. If d is regular, then so is d_0, and has

unit variance function
$$V_0(\xi) = V\left\{f^{-1}(\xi)\right\}\left[f'\left\{f^{-1}(\xi)\right\}\right]^2,$$
where V is the unit variance function of d.

Consider the transformation f defined by
$$f(y) = \int_{y_0}^{y} V^{-\frac{1}{2}}(\mu)\, d\mu, \qquad (1.24)$$
for fixed $y_0 \in \Omega$. This transformation yields a unit deviance d_0 with constant unit variance function $V_0 \equiv 1$. It is called the **variance-stabilizing transformation**. As we shall see below, this transformation stabilizes the asymptotic variance for σ^2 small.

1.5.3 Saddlepoint approximations

We now introduce the saddlepoint approximation for dispersion models, a useful approximation to the probability density function, which is important for the asymptotic theory of generalized linear models.

The **saddlepoint approximation** for a dispersion model with regular unit deviance d is defined for $y \in \Omega$ by
$$p(y;\mu,\sigma^2) \sim \left\{2\pi\sigma^2 V(y)\right\}^{-\frac{1}{2}} \exp\left\{-\frac{1}{2\sigma^2}d(y;\mu)\right\} \quad \text{as} \quad \sigma^2 \to 0, \qquad (1.25)$$
meaning that the ratio of the two sides goes to 1 as σ^2 tends to zero. The saddlepoint approximation is valid for an extensive range of models, and is often very accurate. It is even exact in a few very special cases, such as the normal and simplex distributions. The term 'saddlepoint approximation' comes from the proof in the exponential case (see Chapter 3), but we maintain this terminology for the general case.

The saddlepoint approximation may be interpreted as being half way between the original density and a normal approximation. Thus, if we replace $V(y)$ by $V(\mu)$ in (1.25) and introduce a quadratic approximation to the unit deviance in the exponent of the density, we essentially get the normal approximation considered below in Section 1.5.4. The saddlepoint approximation itself resembles a normal density, if we think of the unit deviance as a squared distance and of $\sigma^2 V(y)$ as the approximate variance of Y.

The saddlepoint approximation fortuitously avoids approximation of the unit deviance in the exponent. This is useful because μ

enters the density only via the unit deviance, so that the saddlepoint approximation preserves the statistical properties regarding inference on μ. The unit deviance is often given by a simple analytic expression, for which no approximation is needed.

On the other hand, the approximation of $a(y; \sigma^2)$ by a function of V in (1.25) is often very practical, because $a(y; \sigma^2)$ tends to be an 'ugly' function, an example being the gamma distribution where $a(y; \sigma^2)$ involves the gamma function. The saddlepoint approximation hence preserves the 'good' parts of the density and gets rid of the 'bad' parts, while having good statistical properties.

Note that the saddlepoint approximation on the right-hand side of (1.25), while positive, is not in general a density function on Ω. However, it may be rescaled to become a density function, motivating the following definition.

Let d be a given regular unit deviance defined on $S \times \Omega$ (not necessarily corresponding to a dispersion model). The corresponding **renormalized saddlepoint approximation** is the density function defined for $y \in \Omega$ by

$$p_0(y; \mu, \sigma^2) = a_0(\mu, \sigma^2) V^{-\frac{1}{2}}(y) \exp\left\{-\frac{1}{2\sigma^2} d(y; \mu)\right\} \quad (1.26)$$

where $a_0(\mu, \sigma^2)$ is a normalizing constant defined by

$$\frac{1}{a_0(\mu, \sigma^2)} = \int_\Omega V^{-\frac{1}{2}}(y) \exp\left\{-\frac{1}{2\sigma^2} d(y; \mu)\right\} dy. \quad (1.27)$$

For exponential dispersion models, (1.26) was studied by Nelder and Pregibon (1987), who called it an **extended quasi-likelihood**, and by Efron (1986), who called it a **double exponential family**.

In the case of a unit deviance d corresponding to a dispersion model, the renormalized saddlepoint approximation provides a second type of approximation to the density function,

$$p(y; \mu, \sigma^2) \sim p_0(y; \mu, \sigma^2) \quad \text{as} \quad \sigma^2 \to 0. \quad (1.28)$$

As an approximation to the density of a dispersion model, the renormalized saddlepoint approximation tends to be slightly more accurate than the ordinary one. The dependence of $a_0(\mu, \sigma^2)$ on μ and the need to calculate the integral (1.27), however, makes the former less useful for some purposes.

While we shall let the term 'renormalized saddlepoint approximation' refer to the density (1.26), the term may also refer to the approximation of a dispersion model density by the approximation (1.28), and it should be clear from the context what is meant. Examples of renormalized saddlepoint approximations will be considered in Chapter 5.

The essence of the saddlepoint approximation is the approximation of the function $a(y; \sigma^2)$. Specifically, the saddlepoint approximation is equivalent to

$$\sigma a(y; \sigma^2) \to \{2\pi V(y)\}^{-\frac{1}{2}} \quad \text{as} \quad \sigma^2 \to 0. \tag{1.29}$$

The saddlepoint approximation is said to be **uniform on compacts** if the convergence in (1.29) is uniform in y on compact subsets of Ω.

We now present the important Laplace approximation, as described in, e.g., Barndorff-Nielsen and Cox (1989, p. 60).

Proposition 1.2 (Laplace approximation) *Define*

$$I(\lambda) = \int_\Omega b(y) e^{\lambda t(y)} \, dy,$$

where Ω is an open interval. Suppose that b is positive, and continuous at $y = \mu$, that t is twice differentiable, has global maximum for $y = \mu \in \Omega$ and that

$$K(\mu) = -t''(\mu) > 0.$$

Then

$$I(\lambda) \sim \sqrt{\frac{2\pi}{\lambda K(\mu)}} b(\mu) e^{\lambda t(\mu)} \quad \text{as} \quad \lambda \to \infty. \tag{1.30}$$

The ordinary and renormalized saddlepoint approximations agree asymptotically for σ^2 small, as we now show, using the Laplace approximation.

Theorem 1.3 *Let d be a given regular unit deviance defined on $C \times \Omega$ (not necessarily corresponding to a dispersion model), and define $a_0(\mu, \sigma^2)$ by (1.27). Then, as $\sigma^2 \to 0$,*

$$a_0(\mu, \sigma^2) \sim \left(2\pi\sigma^2\right)^{-\frac{1}{2}}. \tag{1.31}$$

Proof. Consider the Laplace approximation to the integral (1.27). By the properties of a regular unit deviance, we find that $b(y) = V^{-1/2}(y)$ is continuous, and that $t(y) = -d(y; \mu)/2$ has global maximum for $y = \mu$. Using (1.21) we find $K(\mu) = 1/V(\mu)$. Hence, (1.31) follows from the Laplace approximation (1.30). □

The proof of the saddlepoint approximation for exponential dispersion models is shown in Section 3.5. The following theorem concerns the saddlepoint approximation for proper dispersion models.

Corollary 1.4 *In the case of a proper dispersion model, the saddlepoint approximation is uniform on compacts, and the renormalized saddlepoint approximation is exact.*

Proof. For proper dispersion model we have $a_0(\mu, \sigma^2) = a(\sigma^2)$, making the renormalized saddlepoint approximation exact in this case. Furthermore, (1.31) shows that the ordinary saddlepoint approximation applies to any proper dispersion model, and the continuity of the unit variance function V implies that the saddlepoint approximation is uniform on compacts in this case. □

For the variance-stabilizing transformation f, the density (1.23) of the transformed variable $Z = f(X)$ and its saddlepoint approximation become

$$p_Z(z; \xi, \sigma^2) = a\{f^{-1}(z); \sigma^2\} V^{\frac{1}{2}}(y) \exp\left\{-\frac{1}{2\sigma^2} d_0(z; \xi)\right\}$$
$$\sim (2\pi\sigma^2)^{-\frac{1}{2}} \exp\left\{-\frac{1}{2\sigma^2} d_0(z; \xi)\right\}.$$

Since the unit variance function V_0 for Z is identically one in this case, the Taylor-expansion (1.22) applied to d_0 turns into

$$d_0(\mu_0 + z\sigma; \mu_0 + \mu\sigma) = \sigma^2 (z - \mu)^2 + o(\sigma^2), \quad (1.32)$$

which shows that the saddlepoint approximation is approximately the density of a $N(\mu, \sigma^2)$ distribution for σ^2 small. A formal result on convergence to normality is proved in Theorem 1.5 below.

In the special case of a proper dispersion model, the density of Z becomes

$$p_Z(z; \xi, \sigma^2) = a(\sigma^2) \exp\left\{-\frac{1}{2\sigma^2} d_0(z; \xi)\right\},$$

making ξ the mode point of the density. If $d(y; \mu)$ is monotone as a function of y on each side of μ, the density is unimodal, and becomes more and more peaked and concentrated near μ as σ^2 decreases. In this sense, the interpretation of the parameters μ and σ^2 as position and dispersion parameters, respectively, is especially clear for proper dispersion models.

1.5.4 Convergence to normality

In the next theorem, we show that the saddlepoint approximation implies convergence to normality.

Theorem 1.5 *Let* $Y \sim \mathrm{DM}(\mu_0 + \sigma\mu, \sigma^2)$ *be a reproductive dispersion model with uniformly convergent saddlepoint approximation. Then*

$$\frac{Y - \mu_0}{\sigma} \xrightarrow{d} N\{\mu, V(\mu_0)\} \quad \text{as } \sigma^2 \to 0, \tag{1.33}$$

where $\xrightarrow{d}$ *denotes convergence in distribution.*

Proof. Inserting (1.22) in the probability density function of the variable $Z = (Y - \mu_0)/\sigma$, we obtain

$$\begin{aligned}
p_Z(z; \mu, \sigma^2) &= \sigma p_Y(\mu_0 + \sigma z; \mu_0 + \sigma\mu, \sigma^2) \\
&= \sigma a(\mu_0 + \sigma z; \sigma^2) \exp\left\{-\frac{(z-\mu)^2}{2V(\mu_0)} + o(1)\right\} \\
&\sim \{2\pi V(\mu_0)\}^{-\frac{1}{2}} \exp\left\{-\frac{1}{2V(\mu_0)}(z-\mu)^2\right\},
\end{aligned}$$

where the approximation to $\sigma a(\mu_0 + \sigma z; \sigma^2)$ is obtained by the uniform convergence in (1.29) together with the continuity of V. The density of Z hence converges to the normal $N\{\mu, V(\mu_0)\}$ density function, and this implies convergence in distribution. $\square$

The following analogous result for the case where the distribution of Y follows the renormalized saddlepoint approximation (1.26) is easily shown using Theorem 1.3,

$$\frac{Y - \mu}{\sigma} \xrightarrow{d} N\{0, V(\mu)\} \quad \text{as } \sigma^2 \to 0.$$

A more general result of the form (1.33) holds under regularity conditions, see Exercise 1.12.

We note that the normal approximation implied by Theorem 1.5 has the form

$$\mathrm{DM}(\mu_0 + \sigma\mu, \sigma^2) \approx N\{\mu_0 + \sigma\mu, \sigma^2 V(\mu_0)\},$$

and applies for any $\mu \in \mathbf{R}$. We interpret this as saying that a dispersion model with small dispersion is locally, and not just pointwise, approximated by the family of normal distributions. Here, by 'locally', we mean for a neighbourhood of μ-values.

1.6 Outline of the rest of the book

The present chapter has dealt with just a few general aspects of dispersion models. The remaining chapters will develop the various special cases in more detail.

NOTES

Chapter 2 is an introduction to natural exponential families. The chapter begins with a summary of the properties of moment and cumulant generating functions. Natural exponential families are then introduced along with the deviance and the variance function. Next, the uniqueness theorem for variance functions is proved, followed by Mora's convergence theorem for variance functions and some of its applications. The chapter ends with a result on the asymptotic behaviour of the variance function at a finite extreme of its domain.

Chapter 3 introduces exponential dispersion models as an extension of natural exponential families, with special emphasis on the reproductive and additive cases and their use for continuous and discrete data, respectively. We then discuss convolution, infinite divisibility and additive processes for exponential dispersion models, followed by a review of the six members of Morris' class of models with quadratic variance functions. After this, the saddlepoint approximations for the density and distribution function of exponential dispersion models are discussed, including normal approximations for distributions of residuals.

Chapter 4 introduces the class of Tweedie exponential dispersion models. This is a class of mainly continuous exponential dispersion models closed under scale transformations, which are related to stable distributions and compound Poisson models. We discuss a kind of generalized central limit theorem for exponential dispersion models, in which the Tweedie models appear as limits, along with some examples. Similar results hold for models with exponential variance functions. Finally, we consider a class of Tweedie-Poisson mixtures, suitable for discrete data.

We return to proper dispersion models in Chapter 5. This chapter begins with a more general definition of proper dispersion models and its relation with the simpler definition of the present chapter. We then consider methods for the construction of new proper dispersion models, and present a number of examples. The chapter ends with a discussion of Studentized proper dispersion models, which provide generalizations of Student's t distribution.

1.7 Notes

We have already traced some of the developments of dispersion models and generalized linear models in Section 1.4. Some of the key references to dispersion models are Sweeting (1981) and Jørgensen (1983, 1987b), who introduced and studied dispersion models in their general form, and Jørgensen (1986, 1987a), who introduced the general form of exponential dispersion models and initiated the systematic study of their properties. Proper dispersion models were introduced by Jørgensen (1997).

The saddlepoint approximation for the distribution of sample averages from the exponential family was introduced by Daniels (1954), and has subsequently been generalized and studied extensively, see Reid (1988, 1995) and references therein. It may be considered a special case of *Barndorff-Nielsen's formula* (the p^* formula) for the distribution of the maximum likelihood estimator (Barndorff-Nielsen, 1980, 1983, 1988, 1990).

1.8 Exercises

Exercise 1.1 Make a list of univariate distributions for different data types, classified according to the type of its support. Make sure that you have at least one distribution for each type of support mentioned in the list on p. 2. Determine which, if any, of the distributions on your list are dispersion models. You may want to consult dictionaries of distributions such as Johnson, Kotz and Kemp (1992) and Johnson, Kotz and Balakrishnan (1994, 1995).

Exercise 1.2 (Yokes) Let $\Omega \subseteq \mathbf{R}$ be an interval. A **yoke** is a function $t: \Omega \times \Omega \to \mathbf{R}$, such that for all $y \in \Omega$

$$\sup_{\mu \in \Omega} t(y; \mu) = t(y; y).$$

A yoke that satisfies $t(y; y) = 0$ for all $y \in \Omega$ is called a **normed yoke**.

1. Show that if d is a unit deviance defined on $C \times \Omega$, then $-d$ is a normed yoke on $\Omega \times \Omega$.
2. Show that if t is a yoke, then $\tilde{t}(y;\mu) = t(y;\mu) - t(y;y)$ is a normed yoke.
3. If t is a normed yoke, show that for all $\mu \in \Omega$

$$\sup_{y \in \Omega} t(y; \mu) = 0.$$

4. Show that if t is a normed yoke such that $t(y;\mu) < 0$ for $y \neq 0$, then $-t$ is a unit deviance.

Exercise 1.3 Consider a single observation y from a dispersion model $\mathrm{DM}(\mu, \sigma^2)$. Show that if $y \in \Omega$, then y is the maximum likelihood estimate of μ.

EXERCISES

Exercise 1.4 (Beta distribution) Consider the beta distribution with parameters α, β, with density for $0 < y < 1$ given by

$$\frac{1}{B(\alpha,\beta)} y^{\alpha-1} (1-y)^{\beta-1},$$

where the beta function B is defined by

$$B(\alpha,\beta) = \frac{\Gamma(\alpha)\Gamma(\beta)}{\Gamma(\alpha+\beta)}.$$

1. Define the parameters λ and μ by $\lambda = \alpha + \beta$ and $\mu = \alpha/(\alpha+\beta)$, corresponding to the following form of the density:

$$\frac{1}{B(\alpha,\beta)} y^{-1} (1-y)^{-1} \exp\{\lambda[\mu\log y + (1-\mu)\log(1-y)]\}.$$

Show that this form of the beta distribution is not a dispersion model.

2. Can the beta distribution be reparametrized to become a dispersion model in some other way? Justify your answer.

Exercise 1.5 (t distribution) The t distribution with ν degrees of freedom ($\nu > 0$), denoted $X \sim t(\nu)$, has probability density function

$$p(x;\nu) \frac{1}{\sqrt{\nu}B\left(\frac{\nu}{2},\frac{1}{2}\right)} \left(1 + \frac{x^2}{\nu}\right)^{-\frac{\nu+1}{2}}.$$

Define

$$Y = \frac{X}{\sqrt{\nu}} + \mu,$$

where $\mu \in \mathbf{R}$.

1. Show that Y follows a location-dispersion model with dispersion parameter $\sigma^2 = 1/(\nu+1)$ and unit deviance $d(y-\mu)$ corresponding to the function

$$d(y) = \log(1+y^2).$$

2. Find the unit variance function corresponding to the unit deviance $d(y-\mu)$, and derive the saddlepoint approximation for this location-dispersion model.

3. Show the convergence result

$$X \xrightarrow{d} N(0,1) \quad \text{as} \quad \nu \to \infty,$$

using the normal convergence result for dispersion models.

Exercise 1.6 Make plots of the density functions of the location-dispersion model from Exercise 1.5 model for selected values of μ and σ^2.

Exercise 1.7 (Hyperbolic distribution) Study the hyperbolic distribution, given by the density function

$$\frac{1}{2\alpha K_1(\lambda)} \exp\left(-\lambda \left[\alpha\left\{1 + (y-\mu)^2\right\}^{1/2} - \beta(y-\mu)\right]\right),$$

for $y \in \mathbf{R}$, where $\alpha^2 = 1 + \beta^2$ and K_1 is a Bessel function.

1. Show that this is a location-dispersion model for each fixed value of α, and find the unit deviance and the unit variance function.

2. Use the saddlepoint approximation to derive an asymptotic approximation for $K_1(\lambda)$ for λ large.

Exercise 1.8 (Hyperbola distribution) The hyperbola distribution is defined by the probability density function

$$p(y; \mu, \lambda) = \frac{1}{2K_0(\lambda)} y^{-1} \exp\left\{-\frac{\lambda}{2}\left(\frac{y}{\mu} + \frac{\mu}{y}\right)\right\}, \quad y > 0,$$

where $\lambda > 0$ and $\mu > 0$ are parameters and K_0 is a Bessel function.

1. Show that the hyperbola distribution is a scale-dispersion model with dispersion parameter $1/\lambda$.

2. Find the corresponding unit deviance and unit variance function, and find the saddlepoint approximation.

Exercise 1.9 (Exponential power family) Consider the location-dispersion model defined by

$$d_\rho(y) = |y|^\rho,$$

where $\rho > 0$.

1. Show that the corresponding density function is

$$p_\rho(y; \mu, \sigma^2) = \frac{\rho\left(2\sigma^2\right)^{-1/\rho}}{2\Gamma(1/\rho)} \exp\left\{-\frac{1}{2\sigma^2} |y - \mu|^\rho\right\}, \quad y \in \mathbf{R}.$$

2. Show that this is a location-scale model, and identify the scale parameter.

3. Determine for which values of ρ the unit deviance $d_\rho(y - \mu)$ is regular, and in the affirmative cases find the corresponding unit variance function.

4. Find the saddlepoint approximation to the density function in the cases where the unit deviance is regular.

EXERCISES

Exercise 1.10 (Box-Cox transformation) Consider the transformation f_ρ defined for each $\rho \in \mathbf{R}$ by

$$f_\rho(y) = \begin{cases} \dfrac{y^\rho - 1}{\rho} & \text{for } \rho \neq 0 \\ \log y & \text{for } \rho = 0, \end{cases}$$

for $y > 0$.

1. Find the image of $\mathbf{R}_+$ by this transformation for each $\rho \in \mathbf{R}$.
2. Show that this is the variance-stabilizing transformation for the unit variance function

$$V(\mu) = \mu^{2(1-\rho)}, \quad \mu > 0.$$

3. Find the variance-stabilizing transformation for the Poisson and gamma distributions.

Exercise 1.11 Find the variance-stabilizing transformation for the binomial distribution.

Exercise 1.12 Assume that Y follows a renormalized saddlepoint approximation with parameters $\mu_0 + \sigma\mu$ and σ^2. Define regularity conditions under which the following result holds for $\mu \in \mathbf{R}$:

$$\frac{Y - \mu_0}{\sigma} \xrightarrow{d} N\{\mu, V(\mu_0)\} \quad \text{as } \sigma^2 \to 0.$$

CHAPTER 2

Natural exponential families

In this chapter we consider basic results for moment generating functions, natural exponential families and variance functions. This material prepares the ground for the study of exponential dispersion models in Chapter 3.

2.1 Moment and cumulant generating functions

We begin with a review of some important results for moment and cumulant generating functions needed for natural exponential families. Most of the proofs are omitted, and may be found in, for example, Feller (1971). An elementary introduction to moment generating functions may be found in, for example, Hoel, Port and Stone (1971, Chapter 8).

2.1.1 Definition and properties

We define the **moment generating function** for a random variable Y by

$$M_Y(s) = Ee^{sY}, \qquad (2.1)$$

and the **cumulant generating function** by

$$K_Y(s) = \log M_Y(s).$$

Note that $M_Y(s)$ and $K_Y(s)$ are defined as infinite if the random variable e^{sY} does not have expectation. The set

$$\Theta_Y = \{s \in \mathbf{R} : Ee^{sY} < \infty\}$$

is called the (effective) **domain** of M_Y and K_Y. Note that $0 \in \Theta_Y$, $M_Y(0) = 1$ and $K_Y(0) = 0$.

Example 2.1 (Exponential distribution) The unit exponential distribution with density e^{-x} for $x > 0$ has moment generating

function
$$M(s) = \int_0^\infty \exp(-x + sx)\, dx = (1-s)^{-1}$$
for $s \in \Theta = (-\infty, 1)$. Hence $K(s) = -\log(1-s)$ for $s < 1$. In particular, K is a strictly convex function on Θ. □

We call a moment (cumulant) generating function **degenerate** if $\Theta_Y = \{0\}$. In other words, a degenerate moment generating function is infinite for all values of s except $s = 0$. An example of a distribution with a degenerate moment generating function is the Cauchy, see Exercise 2.3. Similarly, we say that the distribution of Y is **degenerate** if Y is constant, i.e. if $P(Y = c) = 1$ for some constant $c \in \mathbf{R}$; such a random variable has moment generating function $M_Y(s) = e^{sc}$.

In Exercise 2.5 it is shown that if Y is a non-negative random variable, then $\mathbf{R}_- \subseteq \Theta_Y$. Similarly, if Y is bounded then $\Theta_Y = \mathbf{R}$. Here and in the following, we use the notations $\mathbf{R}_+ = (0, \infty)$ and $\mathbf{R}_- = (-\infty, 0)$.

The moment and cumulant generating functions satisfy
$$M_{a+bY}(s) = e^{as} M_Y(bs),$$
$$K_{a+bY}(s) = as + K_Y(bs),$$
for constants a and b. If X and Y are independent random variables we have
$$M_{X+Y}(s) = M_X(s) M_Y(s),$$
$$K_{X+Y}(s) = K_X(s) + K_Y(s).$$

2.1.2 Convexity

We now consider convexity of the cumulant generating function.

Theorem 2.1 (Hölder's inequality) *For any $s, t \in \mathbf{R}$ and $0 \leq \alpha \leq 1$ we have*
$$M_Y \{\alpha s + (1-\alpha) t\} \leq M_Y^\alpha(s) M_Y^{1-\alpha}(t). \tag{2.2}$$
If $0 < \alpha < 1$ and $s \neq t$, the inequality is strict if and only if Y is not degenerate.

Proof. The logarithm is a strictly concave function, and hence for any $a > 0$, $b > 0$ and $0 \leq \alpha \leq 1$ we have
$$\alpha \log a + (1-\alpha) \log b \leq \log \{\alpha a + (1-\alpha) b\}. \tag{2.3}$$

MOMENT AND CUMULANT GENERATING FUNCTIONS 39

Let $c = M_Y(s)$ and $d = M_Y(t)$. If either $c = \infty$ or $d = \infty$, the inequality (2.2) is trivial. If both c and d are finite, let

$$a = e^{sY}/c, \quad b = e^{tY}/d.$$

Inserting this in (2.3) and taking the exponential function on both sides we obtain

$$\frac{\exp\{\alpha sY + (1-\alpha)tY\}}{c^\alpha d^{1-\alpha}} \leq \frac{\alpha e^{sY}}{c} + \frac{(1-\alpha)e^{tY}}{d}. \qquad (2.4)$$

Taking expectations on both sides of (2.4), we obtain (2.2). Equality in (2.2) is obtained if and only if $a = b$ in (2.3), that is, $(s-t)Y = c/d$ with probability 1. This happens if and only if Y is degenerate. □

Corollary 2.2 *The set Θ_Y is an interval.*

Proof. Assume that s and t belong to Θ_Y. We must show that $\alpha s + (1-\alpha)t \in \Theta_Y$ for $0 \leq \alpha \leq 1$. By the definition of Θ_Y we have $M_Y(s) < \infty$ and $M_Y(t) < \infty$. Hence, by Hölder's inequality (2.2) we obtain $M_Y\{\alpha s + (1-\alpha)t\} < \infty$, so that $\alpha s + (1-\alpha)t \in \Theta_Y$. □

Theorem 2.3 *The cumulant generating function K_Y is a convex function on Θ_Y, and strictly convex if and only if Y is not degenerate.*

Proof. By taking logs on both sides of inequality (2.2) we obtain

$$K_Y\{\alpha s + (1-\alpha)t\} \leq \alpha K_Y(s) + (1-\alpha)K_Y(t),$$

which shows the convexity. The condition for strict convexity follows from the condition for strict inequality in Hölder's inequality. □

The degenerate distribution gives an example of a cumulant generating function that is not strictly convex (in fact the only example), having cumulant generating function $K_Y(s) = sc$, $s \in \mathbf{R}$. In this case K_Y is convex, but not strictly convex. At the opposite extreme, we find that for $M_Y(s)$ to be finite for some $s \neq 0$ at least one tail of the density function of Y must be exponentially decreasing. An example where this is not the case is the Cauchy distribution, see Exercise 2.3, which has a degenerate moment generating function, corresponding to $\Theta_Y = \{0\}$. More generally, any distribution with both tails behaving as negative powers has $\Theta_Y = \{0\}$, see Exercise 2.4. Hence further conditions are needed to ensure that a distribution is characterized by its moment generating function, as we shall see in Theorem 2.4 below.

2.1.3 Moments and cumulants

If M_Y is a non-degenerate moment generating function, and if 0 is not an endpoint of Θ_Y, we say that M_Y is a **proper** moment generating function, and similarly for a cumulant generating function. If Y has a proper moment generating function, the derivatives of $M_Y(s)$ at $s = 0$ exist, and by differentiating under the expectation sign we obtain the jth *moment* of Y for $j = 0, 1, \ldots$,

$$\mu_j(Y) = EY^j = M_Y^{(j)}(0), \qquad (2.5)$$

where $M_Y^{(j)}$ denotes the jth derivative of M_Y. In particular $\mu_0(Y) = 1$, $\mu_1(Y) = EY$ and $\mu_2(Y) = EY^2$.

If we expand the exponential function e^{sY} of (2.1) in a Taylor-series and interchange expectation and summation, we obtain the following for a proper moment generating function:

$$\begin{aligned} M_Y(s) &= E\left(\sum_{j=0}^{\infty} \frac{s^j Y^j}{j!}\right) \\ &= \sum_{j=0}^{\infty} \frac{s^j}{j!} \mu_j(Y). \end{aligned}$$

This justifies the terminology 'moment generating function', because the moments are the Taylor-series coefficients of M_Y.

In a similar fashion, if Y has a proper cumulant generating function we define the *cumulants* of Y for $j = 0, 1, \ldots$ by

$$\kappa_j(Y) = K_Y^{(j)}(0). \qquad (2.6)$$

We may then expand K_Y in a Taylor-series with the cumulants as coefficients,

$$K_Y(s) = \sum_{j=0}^{\infty} \frac{s^j}{j!} \kappa_j(Y).$$

Using $K_Y = \log M_Y$, we obtain from (2.6) that $\kappa_0(Y) = 0$,

$$\kappa_1(Y) = \frac{M_Y'(0)}{M_Y(0)} = EY$$

and

$$\begin{aligned} \kappa_2(Y) &= \frac{M_Y''(0) M_Y(0) - M_Y'(0)^2}{M_Y(0)^2} \\ &= EY^2 - E^2Y \\ &= \operatorname{var} Y, \end{aligned}$$

MOMENT AND CUMULANT GENERATING FUNCTIONS

so the first two cumulants $\kappa_1(Y)$ and $\kappa_2(Y)$ are the mean and the variance, respectively. By similar calculations we obtain

$$\kappa_3(Y) = E\left(Y - \mu\right)^3,$$

$$\kappa_4(Y) = E\left(Y - \mu\right)^4 - 3\operatorname{var}^2 Y.$$

The jth **standardized cumulant** of Y is defined for $j = 3, 4, \ldots$ by

$$\gamma_j(Y) = \frac{\kappa_j(Y)}{\kappa_2^{j/2}(Y)}.$$

In particular, the third and fourth standardized cumulants, $\gamma_3(Y)$ and $\gamma_4(Y)$, are known as the **skewness** and **kurtosis** of Y, respectively. Note that the normal distribution has skewness and kurtosis zero.

2.1.4 Characteristic functions

The characteristic function for Y is defined for $t \in \mathbf{R}$ by

$$\varphi_Y(t) = M_Y(it) = Ee^{itY},$$

where $i = \sqrt{-1}$ denotes the complex imaginary unit (see e.g. Feller, 1971, Chapter 15). The characteristic function may be found from the moment generating function by analytic continuation, provided M_Y is non-degenerate.

Uniqueness and inversion formulas

By the uniqueness theorem for characteristic functions we obtain the following uniqueness theorem for moment and cumulant generating functions.

Theorem 2.4 (Uniqueness Theorem) *If Y has a non-degenerate moment (cumulant) generating function, that is $\Theta_Y \neq \{0\}$, the distribution of Y is uniquely determined by the moment generating function M_Y, or by the cumulant generating function K_Y.*

The following Fourier inversion formula for the characteristic function allows us to calculate the probability density function corresponding to a given non-degenerate moment generating function. Suppose that the characteristic function $\varphi_Y(s) = M_Y(is)$ is absolutely integrable. Then Y has a continuous distribution, with probability density function given by

$$p_Y(y) = \frac{1}{2\pi} \int_{-\infty}^{\infty} M_Y(it) e^{-ity}\, dt, \quad y \in \mathbf{R}. \qquad (2.7)$$

In the discrete case, we have the following analogous result. Suppose that the characteristic function $\varphi_Y(s) = M_Y(is)$ is periodic with period 2π. Then Y is discrete with support contained in a translation of the integers $a + \mathbf{Z}$, where $0 \leq a < 1$, and has probability function given by

$$p_Y(z) = \frac{1}{2\pi} \int_{-\pi}^{\pi} M_Y(it) e^{-itz} \, dt, \quad z \in a + \mathbf{Z}. \tag{2.8}$$

Convergence

There is also a parallel for moment generating functions to the continuity theorem for characteristic functions. Here we content ourselves with a version for non-negative variables. Recall that $Y \geq 0$ implies $\Theta_Y \supseteq (-\infty, 0]$, and let $\xrightarrow{d}$ denote convergence in distribution.

Theorem 2.5 (Continuity Theorem) *Let $\{P_n : n = 1, \ldots\}$ be a sequence of non-negative distributions. Let M_n be the moment generating function of P_n. If $P_n \xrightarrow{d} P$ as $n \to \infty$, where P is a non-negative distribution with moment generating function M, then $M_n(s) \to M(s)$ as $n \to \infty$ for $s \leq 0$. Conversely, if $M_n(s) \to M(s)$ as $n \to \infty$ for $s \leq 0$ for some function M, and if $M(s) \to 1$ as $s \to 0$, then M is the moment generating function of a non-negative distribution P and $P_n \xrightarrow{d} P$ as $n \to \infty$.*

Proof. See Feller (1971, p. 431).

2.2 Natural exponential families

We now study the basic properties of natural exponential families. This class includes some important elementary distributions, and provides building blocks for constructing exponential dispersion models (Chapter 3).

2.2.1 Definition

Let ν be a given σ-finite measure on $\mathbf{R}$, and define the function $\kappa(\theta)$ for real θ by

$$\kappa(\theta) = \log \int e^{\theta y} \nu(dy), \tag{2.9}$$

NATURAL EXPONENTIAL FAMILIES

with domain

$$\Theta = \left\{ \theta \in \mathbf{R} : \int e^{\theta y} \nu(dy) < \infty \right\}. \tag{2.10}$$

Let P_θ denote the distribution for the random variable Y defined for measurable sets A by

$$P_\theta(Y \in A) = \int_A \exp\{y\theta - \kappa(\theta)\} \nu(dy), \tag{2.11}$$

and consider the family of distributions

$$\mathcal{P} = \{P_\theta : \theta \in \Theta\}.$$

That P_θ is in fact a probability distribution for each $\theta \in \Theta$ follows from the definition of κ. It is easy to show that the support of the distribution P_θ is independent of θ (Exercise 2.11). Hence, we refer to the common support of the members of the family as simply the **support** of the family. In particular, if one member of the family is degenerate (that is, its support consists of a single point), then all members are degenerate. This leads to the following definition.

Definition 2.1 *The family of distributions $\mathcal{P}$ defined by (2.11) is called a (one-parameter)* **natural exponential family** *if the following two conditions are satisfied:*

1. *The distributions in $\mathcal{P}$ are not degenerate.*
2. *The interval Θ is not degenerate.*

Following standard terminology for exponential families, we call y the **canonical statistic** and θ the **canonical parameter** of the family, and the set Θ is called the **canonical parameter domain**. The function κ is called the **cumulant function** (sometimes called the **cumulant generator**) for the family; this terminology will be justified below. We note that both κ and Θ are determined from the measure ν (by (2.9) and (2.10), respectively); hence we may speak of $\mathcal{P}$ as the natural exponential family **generated by the measure** ν. If ν is a probability distribution Q, say, we speak of the natural exponential family as **generated by the distribution** Q. It is easy to see that a natural exponential family may be generated from any of its members (Exercise 2.12).

Sometimes $e^{\theta y} \nu(dy)$ is called **an exponential tilting** of the measure ν. Hence, up to multiplicative constants, the members of an exponential family are exponential tiltings of each other.

Consider the special case

$$\nu(dy) = c(y)\, dy,$$

where dy denotes either Lebesgue measure (**the continuous case**) or counting measure (**the discrete case**). Then the function $p(\,\cdot\,;\theta)$ defined by
$$p(y;\theta) = c(y)\exp\{\theta y - \kappa(\theta)\} \tag{2.12}$$
is the probability density function of P_θ with respect to Lebesgue measure or counting measure, respectively. A third special case that will be useful later on is the (positive) **mixed case**, where the density is of the form (2.12), but now satisfies
$$c(0) + \int_0^\infty c(y)e^{\theta y}\,dy = \exp\{\kappa(\theta)\}. \tag{2.13}$$
In this case Y has a positive probability mass at zero given by
$$P_\theta(Y=0) = c(0)\exp\{-\kappa(\theta)\},$$
and the distribution is continuous for $y > 0$, such that
$$P_\theta(Y \in A) = \int_A \exp\{y\theta - \kappa(\theta)\}c(y)\,dy \tag{2.14}$$
for $A \subseteq \mathbf{R}_+$ a measurable set. Here dy means Lebesgue measure in both (2.13) and (2.14).

2.2.2 Moment generating functions

We now consider the moment generating function of the members of a natural exponential family. By (2.9), the distribution P_θ has moment generating function
$$\begin{aligned} M(s;\theta) &= \int e^{sy}\exp\{y\theta - \kappa(\theta)\}\,\nu(dy) \\ &= \int \exp\{y(\theta+s)\}\,\nu(dy)\exp\{-\kappa(\theta)\} \\ &= \exp\{\kappa(\theta+s) - \kappa(\theta)\}, \end{aligned} \tag{2.15}$$
for $s \in \Theta - \theta$. Using properties of the moment generating function, we may prove the following result.

Proposition 2.6 *Let $\mathcal{P}$ be the natural exponential family defined by (2.11), with canonical parameter domain Θ. Then*

1. *Θ is an interval.*
2. *Each member of $\mathcal{P}$ is characterized by its moment generating function (2.15).*
3. *$\mathcal{P}$ is parametrized by θ.*

NATURAL EXPONENTIAL FAMILIES   45

Proof. By (2.15), the set $\Theta - \theta$ is the effective domain of the moment generating function $M(\,\cdot\,;\theta)$, which by Corollary 2.2 is an interval. Hence Θ itself is an interval. By Item 2. of the definition, the moment generating function $M(\,\cdot\,;\theta)$ is not degenerate. Hence, by Theorem 2.4, the distribution P_θ is characterized by its moment generating function $M(\,\cdot\,;\theta)$. The proof that $\mathcal{P}$ is parametrized by θ is given in Section 2.2.3. $\square$

The following theorem shows that any moment generating function of the form $\exp\{\kappa(\theta + s) - \kappa(\theta)\}$ corresponds to a natural exponential family.

Theorem 2.7 *Let T be a non-degenerate interval, and let $k(t)$ be a function defined on T, such that for $t, s + t \in T$,*

$$m(s;t) = \exp\{k(t+s) - k(t)\}, \qquad (2.16)$$

is a moment generating function for some non-degenerate distribution P_t. Then the family $\mathcal{P} = \{P_t : t \in T\}$ is a subset of a natural exponential family.

Proof. Since T is non-degenerate, each moment generating function $m(\,\cdot\,;t)$ characterizes the corresponding distribution P_t. Choose $t_0 \in T$, and consider the natural exponential family generated by P_{t_0}. This family has cumulant function κ defined by

$$\begin{aligned}\exp\{\kappa(\theta)\} &= \int e^{\theta y} P_{t_0}(dy) \\ &= m(\theta; t_0) \\ &= \exp\{k(t_0 + \theta) - k(t_0)\}.\end{aligned}$$

The moment generating function of P_{t_0} is

$$\begin{aligned}M(s;\theta) &= \exp\{k(t_0 + \theta + s) - k(t_0) - k(t_0 + \theta) + k(t_0)\} \\ &= \exp\{k(t_0 + \theta + s) - k(t_0 + \theta)\} \\ &= m(s; t_0 + \theta)\end{aligned}$$

for $\theta \in T - t_0$. Hence, each moment generating function $m(s;t) = M(s; t - t_0)$ corresponds to a member of the exponential family generated by P_{t_0}, which proves the theorem. $\square$

Note that if we replace the cumulant function k in (2.16) by $c + k(\,\cdot\, + t_0)$ for constants c and t_0 then (2.16) with T replaced by $T - t_0$ represents the same family. However, the representation (2.16) is unique up to the pair of constants c and t_0.

2.2.3 Moments and cumulants

Consider the distribution P_θ with moment generating function $M(s;\theta) = \exp\{\kappa(\theta+s) - \kappa(\theta)\}$. The ith moment $\mu_i(\theta)$ of P_θ is

$$\mu_i(\theta) = EY^i = M^{(i)}(0;\theta), \quad \theta \in \text{int}\,\Theta.$$

The cumulant generating function for P_θ is

$$\begin{aligned} K(s;\theta) &= \log M(s;\theta) \\ &= \kappa(\theta+s) - \kappa(\theta), \quad s \in \Theta - \theta. \end{aligned}$$

Hence, the ith cumulant $\kappa_i(\theta)$ of P_θ is

$$\kappa_i(\theta) = K^{(i)}(0;\theta) = \kappa^{(i)}(\theta), \quad \theta \in \text{int}\,\Theta. \qquad (2.17)$$

This explains the terminology 'cumulant function' for the function κ.

Expectation

The expectation μ of P_θ is

$$\mu = \kappa'(\theta), \quad \theta \in \text{int}\,\Theta.$$

The function $\tau(\theta) = \kappa'(\theta)$, which gives the relation between the canonical parameter θ and the mean μ is called the **mean value mapping**, and the image $\Omega = \tau(\text{int}\,\Theta)$ is called the **mean domain**. Since τ is continuous and int Θ is an interval, we find that Ω is also an interval. We show in Section 2.3 that τ is a strictly increasing function. Hence, Ω is an open interval, being the image of the open interval int Θ by the strictly increasing and continuous function τ. Figure 2.1 illustrates the mean value mapping and the relation between Θ and Ω; note in particular that for distributions with positive support, Θ is unbounded to the left.

The above results show that μ parametrizes the family $\mathcal{P}_0 = \{P_\theta : \theta \in \text{int}\,\Theta\}$; this is called the **mean value parametrization**. We use the notation $\text{NE}(\mu)$ to denote the distribution in $\mathcal{P}$ with mean μ. If Θ is open, the mean value parametrization parametrizes the whole family $\mathcal{P}$. If Θ contains one or both of its endpoints, the mean value parametrization may be extended by continuity to the endpoints, if necessary by allowing μ to be infinite. In this sense, we may always parametrize a natural exponential family by the mean value parametrization. Since τ is strictly increasing it follows that $\mathcal{P}$ is also parametrized by θ.

NATURAL EXPONENTIAL FAMILIES

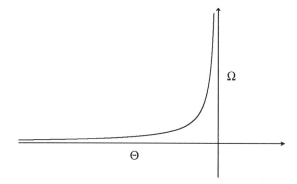

Figure 2.1. *The typical form of the mean value mapping τ, which maps* int Θ *onto* Ω.

2.2.4 Support and steepness

From Exercise 2.11 we know that the support S of a natural exponential family $\mathcal{P}$ is the same for all members of the family. Let C denote the interval from inf S to sup S, with the convention that if an endpoint of C has positive probability, it is included in C, otherwise not. The interval C is the convex support of $\mathcal{P}$, as defined in Section 1.2. The relation between S and the mean domain Ω is the subject of the next theorem.

Lemma 2.8 *Let S be the support of a random variable Y. If Y is not degenerate, then $EY > \inf S$.*

Proof. Let $s_0 = \inf S$, and let P denote the distribution of Y. Then

$$\begin{aligned} EY &= \int_{y \geq s_0} y \, P(dy) \\ &\geq \int_{y \geq s_0} s_0 \, P(dy) = s_0 \int P(dy) = s_0. \end{aligned}$$

The inequality is sharp, because if $EY = s_0$ then $Y = s_0$ almost surely. This shows that $EY > s_0$, proving the theorem. □

Theorem 2.9 *For any natural exponential family we have $\Omega \subseteq$ int C.*

Proof. If $\mu \in \Omega$, then by the lemma, $\mu > \inf S$ because the distribution is not degenerate. Similarly $\mu < \sup S$. Hence $\Omega \subseteq$ int C. □

Definition 2.2 *A natural exponential family* $\mathcal{P}$ *is called* **steep** *if* $\Omega = \text{int}\, C$ *and* **regular** *if the canonical parameter domain* Θ *is open.*

Most of the families that we consider are steep. In many cases, steepness may be shown easily by the following result, due to Barndorff-Nielsen (1978a, p. 117).

Theorem 2.10 (Barndorff-Nielsen) *A regular natural exponential family is steep.*

An example of a family that is not steep is given in Exercise 2.28. We discuss steepness further in Chapter 3. Steepness and regularity are particularly important in connection with maximum likelihood estimation for natural exponential families.

2.3 Variance function and deviance

We now consider the variance function, which plays an important role for both natural exponential families and exponential dispersion models.

The variance of Y, or the second cumulant is given by
$$\operatorname{var} Y = \kappa''(\theta) = \tau'(\theta), \quad \theta \in \text{int}\, \Theta.$$

By Item 1. of Definition 2.1 we have $\operatorname{var} Y > 0$, and hence $\tau'(\theta) > 0$ for $\theta \in \text{int}\, \Theta$. Consequently, τ is a strictly increasing function, a fact that we have already used above.

Since τ has an inverse, τ^{-1}, we may express the variance of P_θ in the form $\operatorname{var}(Y) = V(\mu)$, $\mu \in \Omega$, where the function $V : \Omega \to \mathbf{R}_+$ is defined by
$$V(\mu) = \tau' \left\{ \tau^{-1}(\mu) \right\}.$$

The function V is called the **variance function** for the natural exponential family $\mathcal{P}$. Note that V does not depend on the particular parametrization used in (2.11), because V simply expresses how the variance behaves as a function of the mean μ. We note that $V(\mu) > 0$ for $\mu \in \Omega$.

A useful alternative way of calculating the variance function is
$$V(\mu) = \left\{ \frac{\partial \tau^{-1}}{\partial \mu}(\mu) \right\}^{-1}, \qquad (2.18)$$
which follows by noting that
$$\frac{\partial \tau^{-1}}{\partial \mu}(\mu) = \frac{1}{\tau' \left\{ \tau^{-1}(\mu) \right\}} = \frac{1}{V(\mu)}.$$

VARIANCE FUNCTION AND DEVIANCE

The variance function may also be defined as follows, by the definition of var Y:

$$V(\mu) = \int (y - \mu)^2 \exp\left[y\tau^{-1}(\mu) - \kappa\left\{\tau^{-1}(\mu)\right\}\right] \nu(dy).$$

Deviance

Let $\tilde{\theta}$ be the value of θ that maximizes the function $f(\theta) = y\theta - \kappa(\theta)$. By differentiating f and equating to zero, we obtain the equation

$$y - \tau(\theta) = 0, \tag{2.19}$$

which has solution $\tilde{\theta} = \tau^{-1}(y)$, provided $y \in \Omega$. If $y \in C \setminus \Omega$, the equation (2.19) has no solution, but the maximum of f is usually finite. In the steep case, the set $C \setminus \Omega$ is the boundary of Ω, which has zero probability in the continuous case, but a (usually small) positive probability in the discrete case. In the non-steep case, we have $\tilde{\theta} = \inf \Theta$ if $y < \inf \Omega$ and $\tilde{\theta} = \sup \Theta$ if $y > \sup \Omega$, because, by (2.19), the derivative of f then has constant sign.

Defining

$$l(y; \mu) = y\tau^{-1}(\mu) - \kappa\left\{\tau^{-1}(\mu)\right\},$$

which is the log likelihood for μ, we find that, in the case where $y \in \Omega$, the maximum likelihood estimate for μ is y. In other words, $l(y; \mu)$ is a yoke on $\Omega \times \Omega$. Following the idea in Section 1.2.1, we hence define the (unit) deviance associated with a natural exponential family for $y \in C$ and $\mu \in \Omega$ by

$$d(y; \mu) = 2 \left[\sup_{\theta \in \Theta} \{y\theta - \kappa(\theta)\} - l(y; \mu) \right]. \tag{2.20}$$

For $y, \mu \in \Omega$, the deviance is given by the expression

$$d(y; \mu) = 2 \left[y \left\{\tau^{-1}(y) - \tau^{-1}(\mu)\right\} - \kappa\left\{\tau^{-1}(y)\right\} + \kappa\left\{\tau^{-1}(\mu)\right\} \right].$$

This shows that the deviance is regular (two times continuously differentiable on $\Omega \times \Omega$). For values of y outside Ω, the deviance must be calculated from the definition (2.20).

Consider a natural exponential family generated from a measure ν. The density may then be written in terms of the deviance as follows:

$$p(y; \theta) = a(y) \exp\left\{-\frac{1}{2} d(y; \mu)\right\},$$

where

$$a(y) = c(y) \exp\left[\sup_{\theta \in \Theta} \{y\theta - \kappa(\theta)\}\right].$$

This shows that a natural exponential family may be interpreted as a special case of a dispersion model, having a position parameter μ, but no dispersion parameter ($\sigma^2 = 1$).

For later use, let us consider the first two derivatives of d with respect to μ for $\mu \in \Omega$, which are

$$\frac{\partial d}{\partial \mu}(y; \mu) = -2 \frac{y - \mu}{V(\mu)}$$

and

$$\frac{\partial^2 d}{\partial \mu^2}(y; \mu) = 2 \left\{ \frac{1}{V(\mu)} + (y - \mu) \frac{V'(\mu)}{V^2(\mu)} \right\},$$

where we have used (2.18). In particular we find

$$\frac{\partial^2 d}{\partial \mu^2}(\mu; \mu) = \frac{2}{V(\mu)}.$$

which shows that the variance function for a natural exponential family coincides with the unit variance function for the (unit) deviance d, as defined in Chapter 1. Similarly, the derivative of d with respect to $y \in \Omega$ is

$$\begin{aligned} \frac{\partial d}{\partial y}(y; \mu) &= 2 \left\{ \tau^{-1}(y) + \frac{y}{V(y)} - \frac{y}{V(y)} - \tau^{-1}(\mu) \right\} \\ &= 2\{\tau^{-1}(y) - \tau^{-1}(\mu)\}. \end{aligned}$$

The second derivative is hence

$$\frac{\partial^2 d}{\partial y^2}(y; \mu) = \frac{2}{V(\mu)},$$

which shows that $d(\,\cdot\,; \mu)$ is a convex function on Ω, see also Exercise 2.26.

An alternative expression for the deviance is

$$d(y; \mu) = 2 \int_\mu^y \frac{y - t}{V(t)} \, dt,$$

see Exercise 2.25.

2.3.1 Uniqueness theorem for variance functions

We now show that the variance function characterizes the natural exponential family from which it comes, within the class of all natural exponential families. The role of the variance function for exponential families is hence similar to the role played by the characteristic function or the moment generating function for distributions.

VARIANCE FUNCTION AND DEVIANCE   51

Theorem 2.11 (Uniqueness Theorem) *The variance function V with domain Ω characterizes $\mathcal{P}$ within the class of all natural exponential families.*

Proof. Let $\mathcal{P}_1$ and $\mathcal{P}_2$ be two natural exponential families having the same variance function V with domain Ω. Let τ_i, Θ_i, κ_i, $i = 1, 2$, denote the mean value mapping, canonical parameter domain and cumulant function for $\mathcal{P}_i$, respectively. Then, for $i = 1, 2$,

$$\frac{\partial \tau_i^{-1}}{\partial \mu} = \frac{1}{\tau_i' \left\{ \tau_i^{-1}(\mu) \right\}} = \frac{1}{V(\mu)}, \quad \mu \in \Omega. \tag{2.21}$$

Hence, there exists a constant $\theta_0 \in \mathbf{R}$ such that

$$\tau_1^{-1}(\mu) = \tau_2^{-1}(\mu) + \theta_0, \quad \mu \in \Omega,$$

or, equivalently,

$$\tau_1(\theta) = \tau_2(\theta - \theta_0), \quad \theta \in \operatorname{int} \Theta_1.$$

Since κ_i satisfies the equation

$$\kappa_i'(\theta) = \tau_i(\theta), \quad \theta \in \operatorname{int} \Theta_i,$$

we find that

$$\kappa_1(\theta) = k + \kappa_2(\theta - \theta_0), \quad \theta \in \operatorname{int} \Theta_1,$$

for some constant $k \in \mathbf{R}$. It follows that $\operatorname{int} \Theta_2 = \operatorname{int} \Theta_1 - \theta_0$. Hence, the cumulant generating function for a member of $\mathcal{P}_1$ is

$$\begin{aligned} K_1(s; \theta) &= \kappa_1(\theta + s) - \kappa_1(\theta) \\ &= \kappa_2(\theta - \theta_0 + s) - \kappa_2(\theta - \theta_0) \\ &= K_2(s; \theta - \theta_0), \quad s \in \operatorname{int} \Theta_2 - (\theta - \theta_0). \end{aligned}$$

It follows that the members of $\mathcal{P}_2$ have the same moment generating functions as the members of $\mathcal{P}_1$, so the two families are identical, except perhaps for the endpoints of the canonical parameter domains. However, since both $\mathcal{P}_1$ and $\mathcal{P}_2$ may be generated from a member of their intersection, the two families must be identical, completing the proof. □

If V is a given variance function, the corresponding natural exponential family may in principle be found by going through the steps of the proof of Theorem 2.11, to find the cumulant generating function and, in turn, the characteristic function. The characteristic function may then be inverted, using for example (2.7) or (2.8), to find the probability density function. However, this is not a trivial operation. An example is considered in Chapter 4, where the exponential families corresponding to the variance functions $V(\mu) = \mu^p$ are found. The variance function will be studied further in Chapter 3.

2.3.2 Examples

We now consider three examples of natural exponential families that play important roles as statistical models, including the binomial and Poisson distributions that we have already considered in Chapter 1.

Exponential distribution

The exponential distribution $\text{Ex}(\mu)$ with mean μ has density function of natural exponential family form

$$\begin{aligned} p(y; \mu) &= \mu^{-1} e^{-y/\mu} \\ &= \exp\{\theta y + \log(-\theta)\} \end{aligned}$$

for $y > 0$, where $\theta = -1/\mu < 0$ is the canonical parameter, and the cumulant function is $\kappa(\theta) = -\log(-\theta)$. The mean and variance are

$$\mu = \tau(\theta) = \frac{1}{-\theta}, \quad \text{and} \quad \tau'(\theta) = \frac{1}{(-\theta)^2},$$

respectively. Hence, the variance function is $V(\mu) = \mu^2$, with domain $\Omega = \mathbf{R}_+$. The domain equals the convex support C, so this family is steep. By Theorem 2.11, this is the only natural exponential family with variance function $V(\mu) = \mu^2$ for $\mu > 0$. Figure 2.2 shows the variance functions of the exponential, Poisson and binomial distributions.

Poisson distribution

The Poisson distribution $\text{Po}(\mu)$ with mean μ has probability function

$$\begin{aligned} p(y; \mu) &= \frac{\mu^y}{y!} e^{-\mu} \\ &= \frac{1}{y!} \exp\left(\theta y - e^\theta\right), \quad y \in \mathbf{N}_0, \end{aligned}$$

where $\theta = \log \mu \in \mathbf{R}$. This is a natural exponential family with canonical parameter θ. It has cumulant function $\kappa(\theta) = e^\theta$, mean $\mu = \kappa'(\theta) = e^\theta$ and variance $\tau'(\theta) = e^\theta$. All the cumulants are equal to e^θ. Since the mean and variance are equal, the variance

VARIANCE FUNCTION AND DEVIANCE

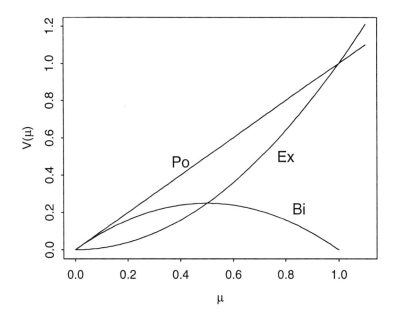

Figure 2.2. *Variance functions for the Poisson, exponential and binomial (with $m = 1$) distributions.*

function is $V(\mu) = \mu$ for $\mu \in \Omega = \mathbf{R}_+$. The support is $S = \mathbf{N}_0$, the convex support is $C = [0, \infty)$ and so the family is steep. Again, by Theorem 2.11, this is the only natural exponential family whose variance function is the identity function on $\mathbf{R}_+$.

Binomial distribution

Let $Y \sim \text{Bi}(m, p)$ be a binomial random variable with probability parameter $p \in (0, 1)$. Then Y has probability function

$$
\begin{aligned}
p^*(y; p, m) &= \binom{m}{y} p^y (1-p)^{m-y} \\
&= \binom{m}{y} \exp\left\{y\theta - m \log\left(1 + e^\theta\right)\right\}
\end{aligned}
$$

for $y = 0, \ldots, m$. This is a natural exponential family for each m with canonical parameter $\theta = \log p/(1-p)$. The mean is $\mu = mp$, and the variance function is $V(\mu) = \mu(1 - \mu/m)$ for $\mu \in (0, m)$.

2.4 Convergence results for variance functions

We now consider two important results on convergence of variance functions.

2.4.1 Convergence of variance functions

Since a natural exponential family is characterized by its variance function, it is natural to ask whether convergence of a sequence of variance functions implies convergence of the corresponding sequence of natural exponential families. We know, for example, that the characteristic function for distributions has such a property. The following theorem, due to Mora (1990), shows that such a convergence result holds for variance functions, provided that the convergence is uniform on compact subsets.

Theorem 2.12 (Mora) *Let* $\{NE_n(\mu)\}_{n=1}^{\infty}$ *denote a sequence of natural exponential families on* $\mathbf{R}$, *with* V_n *being the variance function and* Ω_n *the mean domain of* $NE_n(\mu)$. *Suppose that*

1. $\bigcap_{n=1}^{\infty} \Omega_n$ *contains a non-empty open interval* Ω_0;
2. $\lim_{n \to \infty} V_n(\mu) = V(\mu)$ *exists uniformly on compact subsets of* Ω_0;
3. $V(\mu) > 0$ *on* Ω_0.

Then there exists a natural exponential family $NE(\mu)$, *whose variance function coincides with* V *on* Ω_0, *such that for each* μ *in* Ω_0, *the sequence* $\{NE_n(\mu)\}_{n=1}^{\infty}$ *converges in distribution to* $NE(\mu)$.

Proof. Let $\psi_n = \tau_n^{-1}$ denote the inverse of the mean value mapping for $NE_n(\mu)$. Then $\psi_n'(\mu) = 1/V_n(\mu)$, and similarly, let $\psi'(\mu) = 1/V(\mu)$. Then

$$|\psi_n'(\mu) - \psi'(\mu)| = \frac{|V_n(\mu) - V(\mu)|}{V_n(\mu) V(\mu)}. \qquad (2.22)$$

Let K be a compact subset of Ω_0. By the uniform convergence of $V_n(\mu)$ to $V(\mu)$ on K, it follows that $\{V_n(\mu)\}$ is uniformly bounded on K. Since $V(\mu)$ is bounded on K, it follows from (2.22) and from the uniform convergence of V_n that $\psi_n'(\mu) \to \psi'(\mu)$ uniformly on K.

Next, fix $\mu_0 \in K$ and parametrize each family $NE_n(\mu)$ such that $\psi_n(\mu_0) = 0$. Then, by a result on uniform convergence (Rudin, 1976, p. 152), we find that $\psi_n(\mu) \to \psi(\mu)$ uniformly on K, where $\psi(\mu_0) = 0$. Let $\tau = \psi^{-1}$. Since ψ is monotonically increasing and differentiable, the same is the case for τ.

Now, let $J = \psi(K)$, $\theta = \psi(\mu)$ and $\theta_n = \psi_n(\mu)$ for some $\mu \in K$. Since $V_n(\mu)$ is uniformly bounded on K, there exists an M such that $V_n(\mu) \leq M$ for all n and all $\mu \in K$. Hence we can show that for all $\theta \in J$, $\tau'_n(\theta) \leq M$. Since $\mu = \tau(\theta) = \tau_n(\theta_n)$ we find, using the Mean Value Theorem, that

$$\begin{aligned}
|\tau_n(\theta) - \tau(\theta)| &= |\tau_n(\theta) - \tau_n(\theta_n)| \\
&\leq M |\theta - \theta_n| \\
&= M |\psi(\mu) - \psi_n(\mu)|.
\end{aligned}$$

From this it follows that $\tau_n(\theta) \to \tau(\theta)$ uniformly on J.

Since $\kappa_n(\theta)$ is the cumulant generating function of $\mathrm{NE}_n(\mu_0)$, it follows that $\kappa_n(0) = 0$ for all n. Hence, from the uniform convergence of τ_n, it follows by the result used above that $\kappa_n(\theta)$ converges uniformly on J to a function $\kappa(\theta)$ such that $\kappa(0) = 0$ and $\kappa' = \tau$.

Now, let $[-s_0, s_0]$ denote a neighbourhood where κ_n converges, and let $\nu_n = \mathrm{NE}_n(\mu_0)$. Then for given $a > 0$

$$\begin{aligned}
\nu_n\{(-a,a)^c\} &= \int_{(-a,a)^c} \nu_n(dx) \\
&\leq e^{-s_0 a} \int_{(-a,a)^c} \left(e^{s_0 x} + e^{-s_0 x}\right) \nu_n(dx) \\
&\leq e^{-s_0 a} \left[\exp\{\kappa_n(-s_0)\} + \exp\{\kappa_n(s_0)\}\right].
\end{aligned}$$

It follows that the sequence $\{\nu_n\}$ is tight, so by Helly's Theorem, there exists a subsequence $\{\nu_{n_i}\}$ and a probability measure ν such that $\nu_{n_i} \to \nu$. Since the function $x \mapsto e^{\theta x}$ is continuous we obtain,

$$\int e^{\theta x} \nu_{n_i}(dx) \to \int e^{\theta x} \nu(dx) = \exp\{\kappa(\theta)\},$$

which implies that κ is the cumulant generating function of ν. By the continuity theorem for moment generating functions, it follows that $\nu_n \to \nu$. Obviously the mean of ν is μ_0. Let $\mathrm{NE}(\mu)$ denote the natural exponential family generated by ν. By construction, the variance function for $\mathrm{NE}(\mu)$ is equal to $V(\mu)$ on Ω_0. It is easy to show that $\mathrm{NE}_n(\mu)$ converges to $\mathrm{NE}(\mu)$ for any μ in Ω_0, completing the proof. □

Applications

The next theorem is the first of several important applications of the Mora convergence theorem. This particular application shows that the Central Limit Theorem follows from Mora's Theorem.

For each n, let $X_{1n}, \ldots, X_{nn}$ be independent and identically distributed $\mathrm{NE}(\mu_n)$, where $\mathrm{NE}(\mu)$ denotes a natural exponential family with cumulant function κ and variance function V. Let

$$S_n = X_{1n} + \cdots + X_{nn} \qquad (2.23)$$

and $\mu_n = \tau(\theta_n)$. Then S_n has moment generating function

$$M^n(s; \theta_n) = \exp\{n\kappa(\theta_n + s) - n\kappa(\theta_n)\} \qquad (2.24)$$

and mean $n\mu_n$. By Theorem 2.7 this is the moment generating function of a natural exponential family for each fixed n.

Theorem 2.13 *If $\mu_n = \mu_0 + \mu/\sqrt{n}$ for some $\mu_0 \in \Omega$ and $\mu \in \mathbf{R}$ fixed, then*

$$\frac{S_n - n\mu_0}{\sqrt{n}} \xrightarrow{d} N\{\mu, V(\mu_0)\} \text{ as } n \to \infty. \qquad (2.25)$$

Proof. Define

$$U_n = \frac{S_n - n\mu_0}{\sqrt{n}}$$

and assume that n is so large that $\mu_0 + \mu/\sqrt{n} \in \Omega$. Then, by the result of Exercise 2.13, U_n follows a natural exponential family. The mean and variance of U_n are μ and $V(\mu_0 + \mu/\sqrt{n})$, respectively. The convergence in

$$\lim_{n \to \infty} V\left(\mu_0 + \mu/\sqrt{n}\right) = V(\mu_0)$$

is uniform in μ on compact intervals, due to the continuity of V. The variance function $\mu \mapsto V(\mu_0)$ is constant, corresponding to the natural exponential family $N\{\mu, V(\mu_0)\}$ (see Exercise 2.9), so the conclusion follows from the Mora convergence theorem. □

We note that the theorem is slightly more general than the usual version of the Central Limit Theorem, which corresponds to the special case $\mu = 0$ in (2.25).

The next result shows an example of convergence to the exponential distribution $\mathrm{Ex}(\mu)$.

Proposition 2.14 *Let $NE(\mu)$ denote a natural exponential family with variance function V. If*

$$V(\mu) \sim \mu^2$$

as μ goes to zero or infinity, then

$$c^{-1}\mathrm{NE}(c\mu) \xrightarrow{d} \mathrm{Ex}(\mu) \qquad (2.26)$$

as c goes to zero or infinity, respectively.

CONVERGENCE RESULTS FOR VARIANCE FUNCTIONS   57

Proof. The model on the left-hand side of (2.26) is a natural exponential family for each fixed value of $c > 0$, and the corresponding variance function converges as c tends to either zero or infinity,
$$\frac{1}{c^2} V(c\mu) \to \mu^2.$$
The convergence is easily seen to be uniform in μ on compact intervals, and the result hence follows from the Mora convergence theorem. □

The above results will be generalized in Chapters 3 and 4.

Example 2.2 The uniform distribution on $(0,1)$ generates a natural exponential family $Y \sim \text{NE}(\mu)$ on $(0,1)$ with cumulant function $\kappa(\theta) = \log M(\theta)$, where, for $\theta \neq 0$,
$$M(\theta) = \frac{e^\theta - 1}{\theta},$$
with canonical parameter domain $\Theta = \mathbf{R}$ and mean domain $\Omega = (0,1)$. The cumulant function is, for $\theta < 0$,
$$\kappa(\theta) = -\log(-\theta) + \log(1 - e^\theta).$$

The mean and variance are
$$\mu = \tau(\theta) = \frac{1}{-\theta} - \frac{e^\theta}{1 - e^\theta}, \qquad (2.27)$$
and
$$\kappa''(\theta) = \frac{1}{(-\theta)^2} - \frac{e^\theta}{(1 - e^\theta)^2}, \qquad (2.28)$$
respectively. Since $\theta \to -\infty$ as $\mu \to 0$, equations (2.27) and (2.28) show that $V(\mu) \sim \mu^2$ for μ small. By Theorem 2.14, $c^{-1}\text{NE}(c\mu)$ then converges to the exponential distribution $\text{Ex}(\mu)$ as $c \to 0$. □

2.4.2 Asymptotic behaviour of variance function

Since all members of a natural exponential family share the same variance function, so to speak, the variance function in itself does not characterize any individual member of the family, but only the family as a whole. This raises the question of characterizing properties shared by all members of a natural exponential family in terms of its variance function V. The next result shows a relation between the asymptotic behaviour of the variance function and the asymptotic behaviour of the distributions in the family. This result is important, because it may sometimes allow us to derive information about the asymptotic behaviour of the variance function in order to apply the Mora convergence theorem, even in cases where a closed-form expression for the variance function does not exist.

Let S be the support of the natural exponential family $\mathcal{P}$. In the next theorem we assume that

$$\inf S = 0, \qquad (2.29)$$

so that the family $\mathcal{P}$ is non-negative. We define the distance up to the next point of the support by

$$\delta = \inf [S \setminus \{0\}].$$

Note that $\delta = 0$ for continuous distributions, and $\delta = 1$ for discrete integer-valued distributions. Note also that a distribution with $\delta > 0$ must have a positive probability in zero because of (2.29). In general, δ denotes the distance between zero and the smallest positive point of support of the distribution. The next result is from Jørgensen, Martínez and Tsao (1994); the proof is given below.

Theorem 2.15 (Jørgensen, Martínez and Tsao) *Let $\mathcal{P}$ be a natural exponential family with variance function V and support S. If $\inf S = 0$, then V and Ω satisfy*

1. $\inf \Omega = 0$;
2. $\lim_{\mu \to 0} V(\mu) = 0$;
3. $\lim_{\mu \to 0} V(\mu)/\mu = \delta$;
4. *if $P\{0\} > 0$ then $\lim_{\mu \to 0} V'(\mu) = \delta$;*
5. *if $P\{0\} > 0$ then $\lim_{\mu \to 0} V(\mu)/\mu^2 = \infty$.*

For positive distributions satisfying (2.29), the theorem shows that the left endpoint of Ω is zero and that $V(0^+) = 0$. The main result is 3., which shows that the right derivative of V in zero, denoted $V'(0^+)$, equals δ. Point 4. gives a further interpretation of this result when there is a positive probability in zero, and 5. gives information about the second right derivative of V in zero under the same condition.

It is easy to confirm the theorem in particular examples. For example, the exponential distribution has $\delta = 0$ and the Poisson $\delta = 1$, and the corresponding variance functions clearly satisfy 2. and 3. The Poisson variance function also satisfies 4. and 5. In general, $V'(0^+)$ is positive for discrete distributions and zero for continuous distributions.

CONVERGENCE RESULTS FOR VARIANCE FUNCTIONS 59

Consider the sum S_n from (2.23), with terms having distribution $\text{NE}(\mu_n)$. The next result is an application of Theorem 2.15 to the case where μ_n tends to an endpoint of Ω as n tends to infinity. The theorem, as given here, is essentially the same as a theorem of Pérez-Abreu (1991) for power series distributions. A more general version of the theorem was given by Jørgensen (1986), see Theorem 3.6 in Chapter 3.

Theorem 2.16 *If the family* $\text{NE}(\mu_n)$ *is non-negative, discrete integer-valued, and has positive probabilities in 0 and 1, then by taking* $\mu_n = \mu/n$ *where* $\mu > 0$ *is fixed, we obtain*

$$S_n \xrightarrow{d} \text{Po}(\mu) \text{ as } n \to \infty.$$

Proof. Since the probabilities in 0 and 1 are positive, we have, by Theorem 2.15, that the variance function V satisfies $V'(0^+) = 1$. The mean and variance of S_n are $\mu > 0$ and

$$nV(\mu/n) = \mu \frac{V(\mu/n) - V(0^+)}{\mu/n},$$

respectively. For given $\mu > 0$, this variance function converges to $\mu V'(0^+) = \mu$ as n tends to infinity; the variance function of the Poisson distribution. The convergence is uniform in μ on compact sub-intervals of $(0, \infty)$. Hence, the result follows from Mora's convergence theorem. □

There exist many Poisson convergence results, but the above result is fairly general, yet simple. In particular, Mora's convergence theorem and Jørgensen, Martínez and Tsao's result provides a more elegant proof compared with a direct proof, see Exercise 2.32 or Pérez-Abreu (1991). A more general Poisson convergence result is given in Exercise 2.31.

Example 2.3 Let X_{in} be a Bernoulli variable with probability $\mu_n = \mu/n$, such that S_n follows a binomial distribution with parameters $(n, \mu/n)$. Then by Theorem 2.16, S_n converges to the Poisson distribution with mean μ. This gives the standard Poisson approximation for the binomial distribution. If we instead take $\mu_n = \mu_0 + \mu/\sqrt{n}$ for some $\mu_0 \in (0, 1)$ and n so large that $\mu_n \in (0, 1)$, then by Theorem 2.13, and using that the variance function of the Bernoulli distribution is $\mu(1 - \mu)$, we obtain

$$\frac{S_n - n\mu_0}{\sqrt{n}} \xrightarrow{d} N\{\mu, \mu_0(1 - \mu_0)\} \text{ as } n \to \infty.$$

For $\mu = 0$ this is just the normal approximation to the binomial distribution. □

Proof of Theorem 2.15

To prove Theorem 2.15, we need the following results. Consider the family (2.11) corresponding to the measure ν, assume that $\inf S = 0$, and define

$$M^{(r)}(\theta) = \int_{[0,\infty)} y^r e^{\theta y} \nu(dy), \qquad r = 0, 1, \ldots$$

where we define $\int_{a_1}^{a_2} f(x)\,\nu(dx) = \int_{[a_1,a_2)} f(x)\,\nu(dx)$ for any interval $[a_1, a_2)$ with $a_1 > -\infty$. Note that $\inf \Theta = -\infty$, because S is bounded from below.

Lemma 2.17 *Let $\theta \in \text{int}\,\Theta$, then*

$$\lim_{\theta \to -\infty} M^{(r)}(\theta) = 0 \qquad \text{for} \qquad r > 0 \tag{2.30}$$

$$\lim_{\theta \to -\infty} M^{(0)}(\theta) = \nu\{0\}. \tag{2.31}$$

Proof. Follows from the Monotone Convergence Theorem.

Lemma 2.18 *Let $k, r > 0$, then*

$$\lim_{\theta \to -\infty} \frac{M^{(k+r)}(\theta)}{M^{(k)}(\theta)} = \delta^r \tag{2.32}$$

$$\lim_{\theta \to -\infty} \frac{M_\nu^{(r)}(\theta)}{M_\nu^{(0)}(\theta)} = 0. \tag{2.33}$$

Proof. Let $b > \delta$ and $k \geq 0$ and $r > 0$. If either $k > 0$ or $\nu\{0\} = 0$ we have

$$\frac{M_\nu^{(k+r)}(\theta)}{M_\nu^{(k)}(\theta)} \leq \frac{\int_\delta^b y^{k+r} e^{\theta y} \nu(dy) + \int_b^\infty y^{k+r} e^{\theta y} \nu(dy)}{\int_\delta^b y^k e^{\theta y} \nu(dy)}$$

$$\leq b^r + R(b),$$

where

$$R(b) = \frac{\int_b^\infty y^{k+r} e^{\theta y} \nu(dy)}{\int_\delta^b y^k e^{\theta y} \nu(dy)}.$$

Now put $\theta = \theta_0 - s$ for some $\theta_0 \in \Theta$ and let $p(y) = e^{-sb} - s(y - b)e^{-sb}$ be the tangent to e^{-sy} at $y = b$. Then the convexity of e^{-sy} implies that $e^{-sy} \geq p(y)$, and hence

$$\int_\delta^b y^k e^{(\theta_0 - s)y} \nu(dy) \geq e^{-sb} \int_\delta^b y^k e^{\theta_0 y} \nu(dy)$$

CONVERGENCE RESULTS FOR VARIANCE FUNCTIONS 61

$$+ e^{-sb} \int_\delta^b s(b-y) y^k e^{\theta_0 y} \nu(dy)$$
$$\geq s e^{-sb} \int_\delta^b (b-y) y^k e^{\theta_0 y} \nu(dy).$$

This gives

$$R(b) \leq \frac{e^{-sb} \int_b^\infty y^{k+r} e^{\theta_0 y} \nu(dy)}{s e^{-sb} \int_\delta^b (b-x) y^k e^{\theta_0 y} \nu(dy)} \to 0 \quad \text{as} \quad \theta \to -\infty,$$

which implies that

$$\lim_{\theta \to -\infty} \frac{M^{(k+r)}(\theta)}{M^{(k)}(\theta)} \leq b^r. \tag{2.34}$$

If $\nu\{0\} = 0$, let $k = 0$ in (2.34). Since b was arbitrary, subject to $b > \delta = 0$, (2.34) implies (2.33). Furthermore, if $\nu\{0\} > 0$, Lemma 2.17 implies (2.33). Thus (2.33) is proved.

To prove (2.32), let $k > 0, r > 0$, and note that

$$\frac{M_\nu^{(k+r)}(\theta)}{M_\nu^{(k)}(\theta)} \geq \frac{\int_\delta^b y^{k+r} e^{\theta y} \nu(dy)}{\int_\delta^b y^k e^{\theta y} \nu(dy) + \int_b^\infty y^k e^{\theta y} \nu(dy)}$$
$$\geq \left\{\frac{1}{\delta^r} + R_0(b)\right\}^{-1}, \tag{2.35}$$

where

$$R_0(b) = \frac{\int_b^\infty y^k e^{\theta y} \nu(dy)}{\int_\delta^b y^{k+r} e^{\theta y} \nu(dy)}.$$

Using the same technique as for $R(b)$, we can show that $R_0(b) \to 0$ as $\theta \to -\infty$. Thus (2.35) implies that

$$\lim_{\theta \to -\infty} \frac{M^{(k+r)}(\theta)}{M^{(k)}(\theta)} \geq \delta^r. \tag{2.36}$$

Since (2.34) holds for any $b > \delta$, (2.34) and (2.36) imply (2.32). □

Lemma 2.19 *Let $p > 1$ and $r > 0$. Then*

$$\frac{\{M^{(r)}(\theta)\}^p}{M^{(rp)}(\theta)} \to 0 \quad \text{as} \quad \theta \to -\infty.$$

Proof. Because y^r is zero at zero for $r > 0$, we may assume that $\nu\{0\} = 0$. Now write

$$M^{(r)}(\theta) = \int_0^\infty y^r e^{\theta y} \nu(dy) = \int_0^\infty y^r e^{\frac{\theta y}{p}} e^{\frac{p-1}{p} \theta y} \nu(dy).$$

By Hölder's inequality we obtain

$$M^{(r)}(\theta) \le \left\{\int_0^\infty y^{rp} e^{\theta y}\, \nu(dy)\right\}^{\frac{1}{p}} \left\{\int_0^\infty e^{\theta y}\, \nu(dy)\right\}^{\frac{p-1}{p}}.$$

Hence

$$\frac{\{M^{(r)}(\theta)\}^p}{M^{(rp)}(\theta)} \le \left\{\int_0^\infty e^{\theta y}\, \nu(dy)\right\}^{p-1} \to 0 \quad \text{as} \quad \theta \to -\infty,$$

where we have used Lemma 2.17 to obtain the limit. □

Proof (of Theorem 2.15). To prove 1., that $\inf \Omega = 0$, we first use Lemma 2.18, which gives

$$\mu = \tau(\theta) = \frac{M^{(1)}(\theta)}{M^{(0)}(\theta)} \to 0 \quad \text{as} \quad \theta \to -\infty.$$

Since $\tau(\theta) \ge 0$ for any $\theta \in \Theta$, 1. is proved.

We now prove 2., that $\lim_{\mu \to 0} V(\mu) = 0$. Since $\tau(\theta)$ is a strictly increasing function of θ, $\mu \to 0$ if and only if $\theta \to -\infty$. Now, since

$$V\{\tau(\theta)\} = \kappa''(\theta) = \frac{M^{(2)}(\theta)}{M^{(0)}(\theta)} - \tau^2(\theta),$$

the result follows from Lemma 2.18. In a similar fashion, we obtain

$$\frac{V(\mu)}{\mu} = \frac{M^{(2)}(\theta)}{M^{(1)}(\theta)} - \frac{M^{(1)}(\theta)}{M^{(0)}(\theta)},$$

so 3. follows from Lemma 2.18.

Now consider 4., that $\lim_{\mu \to 0} V'(\mu) = \delta$ if $\nu\{0\} > 0$. We find that

$$V'\{\tau(\theta)\} = \frac{\kappa'''(\theta)}{\kappa''(\theta)},$$

where

$$\kappa'''(\theta) = \frac{M^{(3)}(\theta)}{M^{(0)}(\theta)} - \frac{M^{(1)}(\theta) M^{(2)}(\theta)}{\{M^{(0)}(\theta)\}^2} - 2\tau(\theta)\kappa''(\theta).$$

This gives

$$V'(\tau(\theta)) = \frac{\frac{M^{(3)}(\theta)}{M^{(2)}(\theta)} - \frac{M^{(1)}(\theta)}{M^{(0)}(\theta)}}{1 - \frac{\{M^{(1)}(\theta)\}^2}{M^{(2)}(\theta) M^{(0)}(\theta)}} - 2\tau(\theta).$$

The last term goes to zero by Item 1. Using Lemma 2.18, we obtain

$$\frac{M^{(3)}(\theta)}{M^{(2)}(\theta)} \to \delta \quad \text{and} \quad \frac{M^{(1)}(\theta)}{M^{(0)}(\theta)} \to 0.$$

Since $\nu\{0\} > 0$, Lemmas 2.17 and 2.19 imply that $M^{(0)}(\theta) \to \nu\{0\} > 0$ and

$$\frac{\{M^{(1)}(\theta)\}^2}{M^{(2)}(\theta)M^{(0)}(\theta)} \to 0,$$

which shows 4.

Finally, consider 5., that $\lim_{\mu \to 0} V(\mu)/\mu^2 = \infty$ if $\nu\{0\} > 0$. In this case we obtain

$$\frac{V(\mu)}{\mu^2} = \frac{M^{(2)}(\theta)}{\{M^{(1)}(\theta)\}^2}M^{(0)}(\theta) - 1,$$

so the result follows from Lemmas 2.17 and 2.19. □

2.5 Notes

The history of exponential families dates back to Fisher (1934). Systematic accounts of exponential family theory have been given by for example Lehmann (1983, 1986), Barndorff-Nielsen (1978a), Johansen (1979) and Brown (1986). Interest in exponential families continues to be strong, not least because of generalized linear models and related ideas. The application of differential-geometric ideas in statistics has also stimulated interest in exponential families, starting with the paper by Efron (1975).

The first to have discovered the uniqueness theorem for variance functions seems to be Tweedie (1947), but later Wedderburn (1974) and Morris (1982) rediscovered the result. Special cases of the result are scattered throughout the statistical literature.

The convergence results for variance functions based on the Mora convergence theorem given in this chapter are all special cases of a result due to Jørgensen, Martínez and Tsao (1994), which we consider in detail in Chapter 4. The Central Limit Theorem and Poisson convergence results are classical, but the idea of using the variance function to prove convergence is due to Morris (1982), although the formal proof is due to Mora (1990).

Many Poisson convergence results for discrete random variables are known in probability theory, see Serfling (1978) and Matsunawa (1986), but the development of similar results for distributions that are only partly discrete (Exercise 2.31, from Jørgensen, Martínez and Tsao (1994)) appears to be new.

2.6 Exercises

Exercise 2.1 Find the moment and cumulant generating functions for the standard normal distribution.

Exercise 2.2 Let X be a random variable with moment generating function $M(s)$. Show that $M(s) \geq 1 + sEX$ for $s \in \mathbf{R}$.

Exercise 2.3 (Cauchy distribution) Let

$$f(x) = \frac{1}{\pi(1+x^2)}$$

denote the probability density function of the Cauchy distribution. Show that this distribution has $\Theta_P = \{0\}$, $M_P(0) = 1$ and $M_P(s) = \infty$ for $s \neq 0$.

Exercise 2.4 Consider a probability density function f such that its tails behave as powers of y,

$$f(y) \sim |y|^{p-1} \quad \text{as} \quad |y| \to \infty,$$

where $p < 0$.

1. Show that all such distributions have the same moment generating function.
2. Generalize this to the case where the two tails of f are governed by different powers.

Exercise 2.5 Show that if Y is a non-negative random variable (meaning that $Y \geq 0$ with probability one), then $\mathbf{R}_- \subseteq \Theta_Y$, and if Y is negative, then $\mathbf{R}_+ \subseteq \Theta_Y$. Similarly, show that if Y is bounded then $\Theta_Y = \mathbf{R}$.

Exercise 2.6 Show that the moment generating function is convex, but not necessarily strictly convex. Hint: It may be useful to consider the degenerate distribution.

Exercise 2.7 (Probability generating function) Let Z be a discrete, non-negative, integer-valued random variable, and define the probability generating function of Z by

$$q_Z(u) = Eu^Z.$$

1. Find the relation between the probability generating function and the moment generating function of Z. What is the relation between the domains of the two functions?
2. Express the mean and variance in terms of the probability generating function. Hint: Differentiate q_Z.

3. Find an inversion formula for the probability generating function, expressing the probability function in terms of the probability generating function. Hint: Expand q_Z in a series with the probabilities as coefficients.

Exercise 2.8 Show that the first cumulant satisfies the relation
$$\kappa_1(a+bY) = a + b\kappa_1(Y),$$
and for the higher cumulants,
$$\kappa_j(a+bY) = b^j \kappa_j(Y) \quad \text{for} \quad j \geq 2.$$
Show that the standardized cumulants satisfy, for $b \neq 0$,
$$\gamma_j(a+bY) = \gamma_j(Y) \quad \text{for} \quad j \geq 3.$$

Exercise 2.9 Show that the normal distribution $N(\mu, \sigma_0^2)$ with $\mu \in \mathbf{R}$ and known variance σ_0^2 is a natural exponential family, and show that it has constant variance function $V(\mu) = \sigma_0^2$, for $\mu \in \mathbf{R}$.

Exercise 2.10 Let $\chi^2(f)$ denote the chi-square distribution with f either integer or real.

1. Show that the distribution $a\chi^2(f)$ with $a > 0$ and f known is a natural exponential family, and find its variance function.
2. Show that the distribution $\log \chi^2(f)$ with $f > 0$ is a natural exponential family.

Exercise 2.11 Show that the support of a natural exponential family is independent of the canonical parameter. Hint: Show that the support is
$$S = \left\{ z \in \mathbf{R} : \int_{\omega(z)} \nu(dy) > 0 \quad \forall \omega(z) \right\},$$
where $\omega(z)$ denotes a neighbourhood of z.

Exercise 2.12 Show that a natural exponential family may be generated from any of its members.

Exercise 2.13 Assume that X follows a natural exponential family. Show that the distribution of $Y = a + bX$, $b \neq 0$, follows a natural exponential family, and find the cumulant function and the variance function.

Exercise 2.14 Let κ_i denote the ith cumulant of a natural exponential family.

1. Show that κ_i satisfies the relation
$$\kappa_{i+1} = V(\mu)\frac{\partial \kappa_i}{\partial \mu}, \quad i = 1, 2, \ldots$$

2. Show that the first two derivatives of V are
$$V'(\mu) = \frac{\kappa_3}{V(\mu)} \quad \text{and} \quad V''(\mu) = \frac{\kappa_4}{V^2(\mu)} - \frac{\kappa_3^2}{V^3(\mu)},$$
respectively.

Exercise 2.15 Show that the cumulants of the exponential distribution are $\kappa_i(\theta) = (i-1)!(-\theta)^{-i}$, for $i = 1, \ldots$, and find the standardized cumulants.

Exercise 2.1 Find the standardized cumulants of the Poisson distribution.

Exercise 2.1 (**Scaled Poisson distribution**) The scaled Poisson distribution is defined by a discrete random variable Y with support
$$\{0, c, 2c, \ldots\},$$
such that $Y/c \sim \text{Po}(\mu)$, where $c > 0$ is a constant. Show that the distribution of Y is a natural exponential family with variance function $c\mu$ for $\mu > 0$.

Exercise 2.1 (**Truncated Poisson distribution**) The truncated Poisson distribution is defined by
$$p(z; \psi) = \frac{\psi^z e^{-\psi}}{z!(1 - e^{-\psi})}, \quad z = 1, 2, \ldots$$
where $\psi > 0$ is a parameter.

1. Show that the truncated Poisson distribution is a natural exponential family with cumulant function $\kappa(\theta) = \log\{\exp(e^\theta) - 1\}$, $\theta \in \mathbf{R}$.

2. Find the expectation and variance of the truncated Poisson distribution using the result 1.

3. If Y follows a truncated Poisson distribution, show that $Y - 1$ is approximately Poisson near zero, and approximately Poisson near infinity.

Exercise 2.1 (**Geometric distribution**) Show that the geometric distribution is a natural exponential family, and find its variance function.

Exercise 2.20 (Logarithmic distribution) We define the logarithmic distribution by the following probability function, for $0 < \rho < 1$, and $y = 1, 2, \ldots$:

$$p(y; \rho) = \frac{\rho^y}{-y \log(1-\rho)}.$$

1. Show that this is a natural exponential family, and show that its mean and variance are

$$EY = \frac{\rho}{b(\rho)(1-\rho)}$$

and

$$\operatorname{var} Y = \frac{\rho\left\{1 - \frac{\rho}{b(\rho)}\right\}}{b(\rho)(1-\rho)^2},$$

where $b(\rho) = -\log(1-\rho)$.

2. Find Θ and Ω.
3. Show that $V(1) = 0$, and that $V'(1^+) = 1$.
4. Make a plot of the variance function of the logarithmic distribution, making sure that the features in point 3. are emphasized. Hint: Plot the points $\{\kappa'(\theta), \kappa''(\theta)\}$ for a suitable set of θ-values.

Exercise 2.21 Make a plot of the variance function of the natural exponential family generated by the uniform distribution $U(0, 1)$. Hint: Plot the points $\{\kappa'(\theta), \kappa''(\theta)\}$ for a suitable set of θ-values.

Exercise 2.22 Let X follow a beta distribution with parameters θ_1 and θ_2. Show that for θ_2 known, $Y = \log X$ follows a natural exponential family. Plot its variance function. Can you find other examples of natural exponential families related to the beta distribution?

Exercise 2.23 Let NE(μ) denote a positive natural exponential family with probability density function

$$p(y; \theta) = c(y) \exp\{\theta y - \kappa(\theta)\}, \quad y \in \mathbf{R}_+$$

1. Show that

$$f(y; \theta) = \frac{y}{\tau(\theta)} p(y; \theta)$$

is a new probability density function. This distribution corresponds to the notion of length biased sampling from renewal theory.

2. Show that this is a new natural exponential family, find its cumulant function, and find the mean and variance.

Exercise 2.24 Find the natural exponential family with variance function $V(\mu) = \mu^2$ for $\mu < 0$.

Exercise 2.25 Let $V : \Omega \to \mathbf{R}_+$ be a continuous function defined on the open interval Ω, and define the function $d : \Omega \times \Omega \to \mathbf{R}$ by

$$d(y; \mu) = 2 \int_\mu^y \frac{y-t}{V(t)} \, dt.$$

1. Show that $d(y; \mu) \geq 0$ on $\Omega \times \Omega$, and that $d(y; \mu) = 0$ if and only if $y = \mu$.

2. Show that if V is the variance function of a natural exponential family, then $d(y; \mu)$ is the corresponding deviance.

Exercise 2.26 Let $d(y; \mu)$ denote the deviance for a natural exponential family $\mathrm{NE}(\mu)$. Show that $d(y; \mu)$ is convex as a function of y for any given μ, and strictly convex when the model is steep.

Exercise 2.27 Define the function ψ by $\psi(\mu) = \kappa \left\{ \tau^{-1}(\mu) \right\}$. Show that ψ satisfies $\psi'(\mu) = \mu/V(\mu)$.

Exercise 2.28 Find an example of a non-steep natural exponential family. Hint: Consider a density proportional to $1/(1+x^2)$ for $x > 0$.

Exercise 2.29 (Laplace exponential family) Let us consider the probability density function

$$c(y) = \frac{1}{2} \exp\left\{ -|y| \right\}.$$

1. Find the natural exponential family generated by c.

2. Show that its variance function is

$$V(\mu) = 1 + \mu^2 + \sqrt{1 + \mu^2}.$$

3. Find the cumulants of this family.

Exercise 2.30 Let $x_1 < x_2 < \cdots < x_k$ be given numbers in $\mathbf{R}$, and define the discrete uniform distribution on $\{x_1, \ldots, x_k\}$ by $c(x_i) = 1/k$, $i = 1, \ldots, k$.

1. Find the natural exponential family generated by $c(y)$, and find in particular the cumulant function.

2. Show that the canonical parameter domain is $\Theta = \mathbf{R}$ and that the mean domain is $\Omega = (x_1, x_k)$.

3. Show that $V'(x_1^+) = x_2 - x_1$. Find $V'(x_k^-)$.

Exercise 2.31 Let Y be a random variable with support S such that $\inf S = 0$, and define, for $0 < p < 1$ and $\delta > 0$,
$$X = \begin{cases} 0 & \text{with probability} \quad p \\ \delta + Y & \text{with probability} \quad 1 - p. \end{cases}$$

1. Show that the variance function of the natural exponential family generated by X satisfies $V'(0^+) = \delta$.
2. Generalize Theorem 2.16 to distributions of this type.

[Jørgensen, Martínez and Tsao, 1994]

Exercise 2.32 Prove Theorem 2.16 by using convergence of the moment generating function.

[Jørgensen, 1986]

Exercise 2.33 (Logistic distribution) Consider the logistic distribution with probability density function
$$c(y) = \frac{e^y}{(1+e^y)^2}.$$

1. Show that the corresponding moment generating function is
$$M(s) = \frac{s\pi}{\sin(s\pi)},$$
and find the natural exponential family generated by the logistic distribution.
2. Find Θ and Ω for this family.
3. Find the mean and variance of the family, expressed in terms of the canonical parameter θ.
4. Show that $V(0) = 1$. How does V behave asymptotically at plus or minus infinity?
5. Make a plot of the variance function of the family. Hint: Plot the points $\{\kappa'(\theta), \kappa''(\theta)\}$ for a suitable set of θ-values.

CHAPTER 3

Exponential dispersion models

The class of exponential dispersion models is one of the two main classes of dispersion models, and provides examples of both reproductive and additive dispersion models. It extends ideas from natural exponential families, and includes many standard families of distributions.

3.1 Definition and properties

3.1.1 Additive and reproductive forms

Before defining exponential dispersion models, we return briefly to the definition of natural exponential families. Thus, consider a σ-finite measure ν on $\mathbf{R}$, define the cumulant function κ by

$$\kappa(\theta) = \int e^{\theta z}\, \nu(dz)$$

and define the canonical parameter domain Θ by

$$\Theta = \{\theta \in \mathbf{R} : \kappa(\theta) < \infty\}.$$

Assume that both ν and Θ are not degenerate, so that for θ in Θ the probability measures

$$\exp\{\theta z - \kappa(\theta)\}\, \nu(dz) \qquad (3.1)$$

represent a natural exponential family. Let $\tau(\theta) = \kappa'(\theta)$ denote the mean value mapping and $\Omega = \tau(\text{int}\,\Theta)$ the mean domain.

Given the natural exponential family (3.1), we define the set Λ to be the set of $\lambda > 0$ such that

$$\lambda \kappa(\theta) = \log \int e^{\theta z}\, \nu_\lambda(dz)$$

for some measure ν_λ. Obviously, Λ contains the value 1. The natural exponential family generated by ν_λ is then

$$\exp\{\theta z - \lambda \kappa(\theta)\}\, \nu_\lambda(dz) \qquad (3.2)$$

for $\theta \in \Theta$. The distribution (3.2) is denoted $\mathrm{ED}^*(\theta, \lambda)$.

Definition 3.1 *The family of distributions of $Z \sim \mathrm{ED}^*(\theta, \lambda)$ for $(\theta, \lambda) \in \Theta \times \Lambda$ is called the* **additive exponential dispersion model** *generated by ν. The corresponding family of distributions of $Y = Z/\lambda \sim \mathrm{ED}(\mu, \sigma^2)$, where $\mu = \tau(\theta)$ and $\sigma^2 = 1/\lambda$, is called the* **reproductive exponential dispersion model** *generated by ν. The parameter μ is defined by continuity at the boundaries of Θ, allowing infinite values if necessary.*

The parameter θ is called the **canonical parameter**, λ is called the **index parameter**, and Λ is called the **index set**. The parameter μ is called the **mean value parameter**, and $\sigma^2 = 1/\lambda$ is called the **dispersion parameter**. The convention used in Definition 3.1 for boundary values of Ω is the same as we used in Chapter 2 for natural exponential families. Similar to the unit deviance and unit variance function, we refer to κ as the **unit cumulant function** of the model. In the following $\mathrm{ED}^*(\Theta, \lambda)$ denotes the natural exponential family (3.2) obtained for fixed λ.

In section 3.1.3, we show that additive and reproductive exponential dispersion models are in fact additive and reproductive dispersion models in the sense defined in Chapter 1. When the context is obvious, we often refer to the two cases of exponential dispersion models as **additive models** and **reproductive models**, respectively.

Just as for dispersion models, we refer to the transformation $Y \mapsto Z = Y/\sigma^2$ as the **duality transformation**, because it provides a duality between the additive and reproductive cases. If we apply the inverse duality transformation to (3.2), we obtain the following representation of the distribution $\mathrm{ED}(\mu, \sigma^2)$:

$$\exp[\lambda\{y\theta - \kappa(\theta)\}]\,\bar{\nu}_\lambda(dy), \qquad (3.3)$$

where $\bar{\nu}_\lambda$ denotes ν_λ transformed by the same transformation. Note that (3.2) and (3.3) are both natural exponential families for λ known. It is useful to summarize the additive and reproductive forms as follows. If $Y = Z/\lambda$, $\mu = \tau(\theta)$ and $\sigma^2 = 1/\lambda$, then

$$Y \sim \mathrm{ED}(\mu, \sigma^2) \iff Z \sim \mathrm{ED}^*(\theta, \lambda).$$

Because of the duality transformation, the additive and reproductive forms may to a large extent be developed in parallel.

DEFINITION AND PROPERTIES

Cumulant generating function

We recall that a natural exponential family has a cumulant generating function of the form
$$s \mapsto \kappa(\theta + s) - \kappa(\theta).$$
Since an additive model corresponds to the cumulant function $\lambda \kappa$, the distribution $\mathrm{ED}^*(\theta, \lambda)$ has cumulant generating function
$$K^*(s; \theta, \lambda) = \lambda\{\kappa(\theta + s) - \kappa(\theta)\} \tag{3.4}$$
for $s \in \Theta - \theta$. Hence, the cumulant generating function of $Y = Z/\lambda \sim \mathrm{ED}(\mu, \sigma^2)$ is
$$K(s; \theta, \lambda) = \lambda\{\kappa(\theta + s/\lambda) - \kappa(\theta)\} \tag{3.5}$$
for $s \in \lambda(\Theta - \theta)$. Here and in the following we use the convention that quantities with and without a star refer to the additive and the reproductive case, respectively.

Densities

In the case where the measures ν_λ in (3.2) have densities $c^*(z; \lambda)$ with respect to some fixed measure (typically Lebesgue measure or counting measure), we obtain the following density for the additive model $\mathrm{ED}^*(\theta, \lambda)$:
$$p^*(z; \theta, \lambda) = c^*(z; \lambda) \exp\{\theta z - \lambda \kappa(\theta)\}, \quad z \in \mathbf{R}. \tag{3.6}$$
Similarly, if the measures $\bar{\nu}_\lambda$ have densities $c(y; \lambda)$ with respect to a fixed measure, the reproductive model $\mathrm{ED}(\mu, \sigma^2)$ has density
$$p(y; \theta, \lambda) = c(y; \lambda) \exp[\lambda\{\theta y - \kappa(\theta)\}], \quad y \in \mathbf{R}. \tag{3.7}$$

Parametrizations

Extending the terminology from natural exponential families, we define the **mean value mapping** $\tau : \mathrm{int}\,\Theta \to \mathbf{R}$ and the **mean domain** Ω by
$$\tau(\theta) = \kappa'(\theta) \quad \text{and} \quad \Omega = \tau(\mathrm{int}\,\Theta),$$
respectively. Also, we define the **unit variance function** $V : \Omega \to \mathbf{R}_+$ by
$$V(\mu) = \tau'\{\tau^{-1}(\mu)\}.$$
By differentiating $K^*(s; \theta, \lambda)$ twice with respect to s and setting $s = 0$, we find the mean and variance in the additive case to be
$$EZ = \xi \quad \text{and} \quad \mathrm{var}\,Z = \lambda V(\xi/\lambda),$$

where $\xi = \lambda\tau(\theta) \in \lambda\Omega$. Similarly, we find in the reproductive case

$$EY = \mu \quad \text{and} \quad \operatorname{var} Y = \sigma^2 V(\mu),$$

where $\mu = \tau(\theta) \in \Omega$.

It is convenient to use the notation $\mu = \tau(\theta)$ in both the reproductive and additive cases, so that the mean of the distribution is μ in the reproductive case and $\lambda\mu$ in the additive case. In this way, the present notation extends that of Chapter 1.

As we know, $\mathrm{ED}(\mu, \sigma^2)$ with σ^2 fixed gives a natural exponential family with variance function $\sigma^2 V(\mu)$. Since distinct values of σ^2 give distinct variance functions $\sigma^2 V(\mu)$, it follows by the uniqueness theorem for variance functions that the parameter (μ, σ^2) provides a parametrization of the family in the reproductive case. Note also that the unit variance function V of a reproductive model in effect characterizes the model, because the set of variance functions $\sigma^2 V(\mu)$ for $\sigma^{-2} \in \Lambda$ characterizes the corresponding set of natural exponential families.

In the additive case $\lambda V(\xi/\lambda)$ is the variance function of the natural exponential family $\mathrm{ED}^*(\Theta, \lambda)$ for λ known, and as a rule, different values of λ give different variance functions. The only exception to this rule is given in the following theorem.

Theorem 3.1 *Let $ED^*(\theta, \lambda)$ be an additive model such that the natural exponential families $ED^*(\Theta, \lambda)$ are identical for all $\lambda \in \Lambda$. If the support contains a positive value, then $ED^*(\Theta, \lambda)$ is a scaled Poisson distribution.*

Proof. If the two natural exponential families $\mathrm{ED}^*(\Theta, \lambda)$ and $\mathrm{ED}^*(\Theta, 1)$, say, are identical, they have the same mean domains and the same variance functions. In terms of the mean domains, this gives

$$\lambda\Omega = \Omega \quad \forall \lambda \in \Lambda \tag{3.8}$$

and for the unit variance functions

$$\lambda V(\mu/\lambda) = V(\mu) \quad \forall \lambda \in \Lambda, \mu \in \lambda\Omega. \tag{3.9}$$

By (3.8) we find that either $\Omega = \mathbf{R}$ or, up to a change of sign, $\Omega = \mathbf{R}_+$. We show in Section 3.2.2 that the index set is unbounded above. By rewriting (3.9) as $V(\mu/\lambda) = V(\mu)/\lambda$ and letting $\lambda \to \infty$, we find that $V(0^+) = 0$, which rules out $\Omega = \mathbf{R}$, because $V(\mu)$ is positive on Ω. Now, using (3.9) again for some $\mu > 0$, we obtain

$$V(\mu) = \frac{V(\mu/\lambda) - V(0)}{\mu/\lambda}\mu \underset{\lambda\to\infty}{\to} V'(0^+)\mu.$$

DEFINITION AND PROPERTIES 75

This shows that the variance function is proportional to μ, which, by Exercise 2.17, implies that the distribution is a scaled Poisson distribution. □

This result shows that, except for the Poisson case, the parameter (θ, λ) provides a parametrization of an additive model. In Section 3.3.3 we show that the Poisson distribution parametrized as $\text{Po}(\lambda e^\theta)$ may be considered an additive exponential dispersion model, providing the exception characterized in the theorem.

Convex support

The following is an immediate corollary to Theorem 2.9 and the definition of steepness from Chapter 2.

Corollary 3.2 *Let C_λ denote the convex support of the natural exponential family $ED(\mu, 1/\lambda)$ for λ known. Then $\Omega \subseteq \text{int}\, C_\lambda$. Moreover, if one such family is steep, then they all are, and $\text{int}\, C_\lambda = \Omega$ for all $\lambda \in \Lambda$.*

The last equation shows that the convex support for a steep reproductive model depends little on λ, or not at all. For an additive model, we obtain the formula

$$\lambda \Omega \subseteq \lambda \,\text{int}\, C_\lambda,$$

with identity if the model is steep.

Let us define the **convex support** of a reproductive model as

$$C = \bigcup_{\lambda \in \Lambda} C_\lambda.$$

The convex support of an additive model is defined as λC. Note that the latter generally depends on λ, except for the cases $C = \pm \mathbf{R}_+$, $C = \pm \mathbf{R}_0$ and $C = \mathbf{R}$.

3.1.2 Continuous and discrete models

We now consider some results that are crucial for the understanding of discrete and continuous exponential dispersion models. We use the terminology **lattice distribution** to denote a distribution whose support is contained in $a + b\mathbf{Z}$ for constants a, b with $b \neq 0$.

Proposition 3.3 *Let $ED^*(\theta, \lambda)$ be an additive model. Then*

1. *If $ED^*(\theta, \lambda)$ is a lattice distribution for some value of θ and λ, then it is lattice with the same b for all $(\theta, \lambda) \in \Theta \times \Lambda$.*
2. *If $ED^*(\theta, \lambda)$ is continuous for some value λ_0 of λ, then it is continuous for any $\lambda \geq \lambda_0$ in Λ.*

Proof. Let us consider the characteristic function of $\mathrm{ED}^*(\theta, \lambda)$,

$$\varphi(t; \theta, \lambda) = \exp\left[\lambda\{\kappa(\theta + it) - \kappa(\theta)\}\right],$$

for $t \in \mathbf{R}$. To prove Item 1. note that $\mathrm{ED}^*(\theta, \lambda)$ is a lattice distribution if and only if the function $|\varphi(\,\cdot\,; \theta, \lambda)|$ is periodic with period $2\pi b$. Since

$$|\varphi(t; \theta, \lambda)| = |\varphi(t; \theta, 1)|^\lambda,$$

we find that if $|\varphi(t; \theta, 1)|$ is periodic, then so is $|\varphi(t; \theta, \lambda)|$, and with the same period. This shows Item 1.

By Fourier inversion, we know that if $\varphi(t; \theta, \lambda_0)$ is absolutely integrable for some $\lambda_0 \in \Lambda$, then $\mathrm{ED}^*(\theta, \lambda_0)$ is continuous. But then

$$|\varphi(t; \theta, \lambda)| = |\varphi(t; \theta, \lambda_0)|^{\lambda/\lambda_0},$$

is also absolutely integrable for any $\lambda \geq \lambda_0$ in Λ. Hence $\mathrm{ED}^*(\theta, \lambda)$ is continuous for $\lambda \geq \lambda_0$ in Λ. This concludes the proof. $\square$

This result has important consequences for the application of exponential dispersion models to discrete data, because it rules out the possibility of integer-valued reproductive models. Indeed, suppose that the distribution $\mathrm{ED}(\mu, 1) = \mathrm{ED}^*(\theta, 1)$ has support contained in $\mathbf{Z}$. Then, by Proposition 3.3, $\mathrm{ED}^*(\theta, \lambda)$ is a lattice distribution with $b = 1$ for any $\lambda \in \Lambda$. By the duality transformation, this implies that the support of $\mathrm{ED}(\mu, \sigma^2)$ is contained in a translation of the set $\{0, \pm\sigma^2, \pm 2\sigma^2, \ldots\}$. Except in pathological cases, the support of $\mathrm{ED}(\mu, \sigma^2)$ is hence not contained in $\mathbf{Z}$. It follows that reproductive models for integer-valued data do not exist, except if σ^2 is taken to be known (giving a natural exponential family). In this sense, the only exponential dispersion models suitable for discrete integer-valued data are additive models.

Item 2. of Proposition 3.3 shows that if a member of an exponential dispersion model is continuous, then the members will be continuous from some value λ_0 of λ upwards. As a rule, the model will be continuous for all values of λ, but the proposition itself does not say anything about what happens for $\lambda < \lambda_0$, except that Item 1. rules out integer-valued distributions. We note that the duality transformation and its inverse transform a continuous model into a continuous one, so that continuous distributions appear in both the additive and the reproductive cases.

Note that, for theoretical purposes, we shall continue to change freely between the additive and reproductive cases by means of the duality transformation, irrespective of the type of distribution.

DEFINITION AND PROPERTIES 77

As an aside, we now use the characteristic function to prove that negative values of λ are impossible, justifying the assumption $\Lambda \subseteq \mathbf{R}_+$ made in connection with (3.2). First, note that φ, being a characteristic function, satisfies $|\varphi(t;\theta,\lambda)| \leq 1$. Hence

$$|\varphi(t;\theta,\lambda)| = |\exp\{\kappa(\theta+it) - \kappa(\theta)\}|^\lambda \leq 1. \tag{3.10}$$

For $\lambda = 1$, we find $|\exp\{\kappa(\theta+it)-\kappa(\theta)\}| \leq 1$, with strict inequality for some $t \in \mathbf{R}$ because the distribution is not degenerate. By (3.10), this implies $\lambda > 0$.

3.1.3 Unit deviance

Consider a reproductive model with unit cumulant function κ and convex support C. By analogy with the definition for natural exponential families in Chapter 2, we define the **unit deviance** of a reproductive model for $y \in C$ and $\mu \in \Omega$ by

$$d(y;\mu) = 2\left[\sup_{\theta \in \Theta}\{y\theta - \kappa(\theta)\} - y\tau^{-1}(\mu) + \kappa\left\{\tau^{-1}(\mu)\right\}\right].$$

Note that this is the deviance of the natural exponential family $\mathrm{ED}^*(\Theta, 1)$. In particular, for $y \in \Omega$ we obtain the expression

$$d(y;\mu) = 2\left[y\left\{\tau^{-1}(y) - \tau^{-1}(\mu)\right\} - \kappa\left\{\tau^{-1}(y)\right\} + \kappa\left\{\tau^{-1}(\mu)\right\}\right].$$

By differentiating d with respect to μ and letting $y = \mu$, we obtain

$$\frac{\partial^2 d}{\partial \mu^2}(\mu;\mu) = \frac{2}{V(\mu)},$$

so again our definition of the unit variance function for exponential dispersion models coincides with the one given in Chapter 1 for arbitrary unit deviances.

The density (3.7) for a reproductive model $Y \sim \mathrm{ED}(\mu,\sigma^2)$ may be written in standard form as follows:

$$p(y;\mu,\sigma^2) = a(y;\sigma^2)\exp\left\{-\frac{1}{2\sigma^2}d(y;\mu)\right\},$$

where

$$a(y;\sigma^2) = c(y;\sigma^{-2})\exp\left[\sigma^{-2}\sup_{\theta \in \Theta}\{y\theta - \kappa(\theta)\}\right].$$

This shows that a reproductive exponential dispersion model is, in fact, a reproductive dispersion model in the sense defined in Chapter 1, justifying the terminology 'reproductive exponential dispersion model'.

For an additive model, reparametrizing from μ to $\xi = \mu/\sigma^2$, the standard form of the density function is

$$p^*(z;\xi,\sigma^2) = a^*(z;\sigma^2)\exp\left\{-\frac{1}{2\sigma^2}d(z\sigma^2;\xi\sigma^2)\right\},$$

where

$$a^*(z;\sigma^2) = c^*(z;\sigma^{-2})\exp\left[\sigma^{-2}\sup_{\theta\in\Theta}\{z\theta - \kappa(\theta)\}\right].$$

This gives an additive dispersion model, justifying the terminology 'additive exponential dispersion model'.

3.1.4 Convergence results

We now consider some important convergence results related to the asymptotic behaviour of the unit variance function. All of these results are special cases of a general convergence theorem for exponential dispersion models to be discussed in Chapter 4. Section 3.3 provides some examples of these convergence results. In the results, we use the fact that the Poisson is a natural exponential family, and that the gamma and normal distributions are natural exponential families for fixed value of the dispersion parameter.

Normal convergence

The following result, which is a version of Theorem 1.5, shows that any exponential dispersion model is asymptotically normal for σ^2 small.

Theorem 3.4 *If $Y \sim \mathrm{ED}(\mu_0 + \sigma\mu, \sigma^2)$ for some $\mu_0 \in \Omega$ and $\mu \in \mathbf{R}$, we have*

$$\frac{Y - \mu_0}{\sigma} \xrightarrow{d} N\{\mu, V(\mu_0)\} \text{ as } \sigma^2 \to 0, \qquad (3.11)$$

where V is the unit variance function of the model.

Proof. Define

$$Y_0 = \frac{Y - \mu_0}{\sigma}$$

and assume that σ^2 is so small that $\mu_0 + \sigma\mu \in \Omega$. Then for σ^2 known, Y_0 follows a natural exponential family with mean μ and variance function $V(\mu_0 + \sigma\mu)$. Since the convergence in

$$\lim_{n\to\infty} V(\mu_0 + \sigma\mu) = V(\mu_0)$$

is uniform in μ on compact intervals, and the variance function $\mu \mapsto$

DEFINITION AND PROPERTIES

$V(\mu_0)$ is constant, corresponding to the natural exponential family $N\{\mu, V(\mu_0)\}$, the conclusion follows from the Mora convergence theorem. □

Taking $\mu = 0$ in the above result, we obtain

$$\frac{Y-\mu}{\sigma} \xrightarrow{d} N\{0, V(\mu)\} \quad \text{as} \quad \sigma^2 \to 0.$$

The corresponding result for additive models is as follows. If $Z \sim \mathrm{ED}^*(\theta, \lambda)$, then for $\theta \in \mathrm{int}\,\Theta$ fixed,

$$\sqrt{\lambda}\left\{\frac{Z}{\lambda} - \tau(\theta)\right\} \xrightarrow{d} N\{0, \tau'(\theta)\} \quad \text{as} \quad \lambda \to \infty. \tag{3.12}$$

Gamma convergence

The following gamma convergence result generalizes Proposition 2.14 from Chapter 2.

Theorem 3.5 *Let* $\mathrm{ED}(\mu, \sigma^2)$ *be a reproductive model with unit variance function* V *and mean domain* $\Omega = \mathbf{R}_+$, *satisfying the asymptotic relation*

$$V(\mu) \sim c_0 \mu^2$$

for $c_0 > 0$, *as* μ *tends to either zero or infinity. Then*

$$c^{-1}\mathrm{ED}(c\mu, \sigma^2) \xrightarrow{d} \mathrm{Ga}(\mu, c_0\sigma^2), \tag{3.13}$$

for c tending to zero or infinity, respectively.

Proof. The left-hand side of (3.13) is a natural exponential family for fixed values of $c > 0$ and $\sigma^{-2} \in \Lambda$, and the corresponding unit variance function converges as c tends to either zero or infinity,

$$\frac{\sigma^2}{c^2}V(c\mu) \to c_0\sigma^2\mu^2.$$

The limiting variance function is that of the gamma natural exponential family on the right-hand side of (3.13). The convergence is easily seen to be uniform in μ on compact intervals, and the result hence follows from the Mora convergence theorem. □

Poisson convergence

The following theorem, which generalizes Theorem 2.16 from Chapter 2, shows that discrete additive exponential dispersion models are approximately Poisson for large λ.

Theorem 3.6 *Let $ED^*(\theta, \lambda)$ denote a non-negative integer-valued additive model with positive probabilities in 0 and 1. Then for $\xi > 0$,*

$$\text{ED}^* \left\{ \tau^{-1}(\xi/\lambda), \lambda \right\} \xrightarrow{d} \text{Po}(\xi) \quad \text{as} \quad \lambda \to \infty.$$

Proof. For λ known, the model $\text{ED}^* \left\{ \tau^{-1}(\xi/\lambda), \lambda \right\}$ is a natural exponential family with mean ξ and variance function $\lambda V(\xi/\lambda)$. Since the probabilities in 0 and 1 are positive, we obtain, using Theorem 2.15 in the case $\lambda = 1$, that V satisfies $V(0^+) = 0$ and $V'(0^+) = 1$. Hence

$$\lambda V(\xi/\lambda) = \xi \frac{V(\xi/\lambda) - V(0^+)}{\xi/\lambda} \to \xi \quad \text{as} \quad \lambda \to \infty,$$

and the convergence is uniform in ξ on compact subsets of $(0, \infty)$. The limiting variance function is that of the Poisson distribution, and the result hence follows from Mora's convergence theorem. □

3.2 Convolution and additive processes

3.2.1 Convolution formula

We now derive a convolution formula for exponential dispersion models. The formula has implications in terms of infinite divisibility and an associated stochastic process, providing an important interpretation of exponential dispersion models.

Additive form

Consider an additive model $\text{ED}^*(\theta, \lambda)$ and assume that $Z_1, \ldots, Z_n$ are independent and

$$Z_i \sim \text{ED}^*(\theta, \lambda_i),$$

for $(\theta, \lambda_i) \in \Theta \times \Lambda$, $i = 1, \ldots, n$. Then the distribution of $Z_+ = Z_1 + \cdots + Z_n$ is

$$Z_+ \sim \text{ED}^*(\theta, \lambda_1 + \cdots + \lambda_n). \tag{3.14}$$

We call this the **additive form** of the convolution formula. The formula shows that an additive model is closed under convolution of members with a common value of the canonical parameter θ, lending support for the terminology 'additive exponential dispersion model'.

The proof of the formula is simple using the cumulant generating function. By the independence of $Z_1, \ldots, Z_n$, and using the fact

CONVOLUTION AND ADDITIVE PROCESSES

that Z_i has cumulant generating function
$$K^*(s;\theta,\lambda_i) = \lambda_i\{\kappa(\theta+s) - \kappa(\theta)\},$$
we find that the cumulant generating function of Z_+ is
$$\begin{aligned} K_{Z_+}(s;\theta,\lambda_1,\ldots,\lambda_n) &= \sum_{i=1}^n \lambda_i\{\kappa(\theta+s) - \kappa(\theta)\} \\ &= K^*(s;\theta,\lambda_1+\cdots+\lambda_n), \end{aligned}$$
which proves (3.14).

Reproductive form

The corresponding formula for reproductive models is easily derived by the duality transformation. Assume that $Y_1,\ldots,Y_n$ are independent and
$$Y_i \sim \mathrm{ED}\left(\mu, \frac{\sigma^2}{w_i}\right), \quad i=1,\ldots,n, \tag{3.15}$$
where $\mu \in \Omega$ and $w_i/\sigma^2 \in \Lambda$ for all i. Letting $w_+ = w_1 + \cdots + w_n$, the **reproductive form** of the convolution formula is
$$\frac{1}{w_+}\sum_{i=1}^n w_i Y_i \sim \mathrm{ED}\left(\mu, \frac{\sigma^2}{w_+}\right). \tag{3.16}$$

The formula shows that a reproductive model is closed under the operation of weighted averaging of members with the same mean, with weights given by the inverse dispersion parameters.

The reproductive form of the convolution formula is easily seen to be equivalent to the additive form (3.14). Thus, by the duality transformation we have
$$\frac{w}{\sigma^2}\mathrm{ED}\left\{\tau(\theta), \frac{\sigma^2}{w}\right\} = \mathrm{ED}^*\left(\theta, \frac{w}{\sigma^2}\right),$$
where $\mu = \tau(\theta)$. If we apply this result to both (3.15) and (3.16), we find that (3.16) is a special case of (3.14) corresponding to $\lambda_i = w_i/\sigma^2$ for $i=1,\ldots,n$.

As a special case of (3.16) we find that for $Y_1,\ldots,Y_n$ independent and identically distributed with distribution $\mathrm{ED}(\mu,\sigma^2)$, the sample average of $Y_1,\ldots,Y_n$ has distribution
$$\frac{1}{n}\sum_{i=1}^n Y_i \sim \mathrm{ED}\left(\mu, \frac{\sigma^2}{n}\right).$$

A family such as this, where the distribution of the sample average belongs to the family itself, is called **reproductive**, lending support to the terminology 'reproductive exponential dispersion model'.

3.2.2 Infinite divisibility

A subject closely related to convolution is infinite divisibility. Let X be a given random variable. If, for a given $n \in \mathbf{N}$, there exist independent and identically distributed random variables $X_1, \ldots, X_n$, such that
$$X = X_1 + \cdots + X_n, \tag{3.17}$$
then X is said to be **divisible** by n. If X is divisible by n for all $n \in \mathbf{N}$, then X is said to be **infinitely divisible**.

Theorem 3.7 *Consider an exponential dispersion model with index set Λ. Then*

1. *The index set Λ is an additive semigroup and $\mathbf{N} \subseteq \Lambda \subseteq \mathbf{R}_+$.*
2. *The members of the model are infinitely divisible if and only if $\Lambda = \mathbf{R}_+$.*

Proof. The convolution formula (3.14) implies that if $\lambda_1, \lambda_2 \in \Lambda$ then $\lambda_1 + \lambda_2 \in \Lambda$, which shows that Λ is an additive semigroup. Because we assume that $1 \in \Lambda$, this implies in particular that $\mathbf{N} \subseteq \Lambda$.

We now prove Item 2. Let $\lambda \in \Lambda$, let $X \sim \mathrm{ED}^*(\theta, \lambda)$ and let n be such that $\lambda/n \in \Lambda$. Then (3.14) gives
$$\mathrm{ED}^*(\theta, \lambda) = \mathrm{ED}^*(\theta, \lambda/n) + \cdots + \mathrm{ED}^*(\theta, \lambda/n). \tag{3.18}$$
Now, if $\Lambda = \mathbf{R}_+$, then $\lambda/n \in \Lambda$ for any $n \in \mathbf{N}$, which shows that (3.17) is satisfied for any $n \in \mathbf{N}$, and hence that $\mathrm{ED}^*(\theta, \lambda)$ is infinitely divisible.

On the other hand, assume that $\mathrm{ED}^*(\theta, \lambda)$ is infinitely divisible, so that (3.18) holds for any n. This shows that $\lambda/n \in \Lambda$ for any $n \in \mathbf{N}$. Since Λ is an additive semigroup, we hence find that $\lambda k/n \in \Lambda$ for any $k, n \in \mathbf{N}$, which, by the duality lemma, we may express as saying that $\mathrm{ED}(\mu, r\sigma^2)$ exists for any positive rational number $r = n/k$. If $\sigma_0^2 > 0$ is given, there exists a sequence of rationals r_i such that $r_i \sigma^2$ tends to σ_0^2, because the rationals are dense in $\mathbf{R}_+$. Hence, we have the following convergence of variance functions:
$$r_i \sigma^2 V(\mu) \to \sigma_0^2 V(\mu),$$
and the convergence is uniform on compact intervals. By the Mora

CONVOLUTION AND ADDITIVE PROCESSES 83

convergence theorem, this proves the existence of $\mathrm{ED}(\mu, \sigma_0^2)$, and since $\sigma_0^2 > 0$ was arbitrary, it follows that $\Lambda = \mathbf{R}_+$. This concludes the proof. □

Item 1. of the theorem shows that the index set Λ is bracketed by the two extremes $\Lambda = \mathbf{N}$ and $\Lambda = \mathbf{R}_+$, where the latter corresponds to infinite divisibility. In particular, Λ is unbounded above. Most examples of exponential dispersion models have either $\Lambda = \mathbf{R}_+$ or $\Lambda = \mathbf{N}$. The infinitely divisible case is the most interesting one from a statistical point of view, because it makes the domain of λ an interval, facilitating inference on λ. We show in Exercise 3.8 that distributions with bounded support (for example on the unit interval) cannot be infinitely divisible, in which case no exponential dispersion models with $\Lambda = \mathbf{R}_+$ exist.

3.2.3 Additive processes

The additive form of the convolution formula for exponential dispersion models is associated with a stochastic process with stationary and independent increments, which provides an important interpretation of the models. In Section 3.3, we show that many familiar examples of stochastic processes are of this form.

Discrete time

Let $\mathrm{ED}^*(\theta, \lambda)$ denote an additive family with $\Lambda \supseteq \mathbf{N}$, and let $\{Z_n : n \geq 0\}$ denote a discrete time stochastic process with $Z_0 = 0$. Define the increments of the process by

$$\Delta Z_n = Z_n - Z_{n-1}.$$

Let us assume that the increments $\Delta Z_1, \Delta Z_2, \ldots$ are independent and identically distributed with distribution

$$\Delta Z_n \sim \mathrm{ED}^*(\theta, 1), \quad n = 1, 2, \ldots \tag{3.19}$$

This is a random walk, which we call a **discrete time additive process**.

The additive form of the convolution formula gives the following distribution for the increments of the process over t steps, for $t = 1, 2, \ldots$:

$$Z_{n+t} - Z_n \sim \mathrm{ED}^*(\theta, t). \tag{3.20}$$

In particular, we have

$$Z_n \sim \mathrm{ED}^*(\theta, n), \quad n = 1, 2, \ldots$$

The mean and variance of the process both increase linearly with time,

$$EZ_n = n\xi \quad \text{and} \quad \text{var}\, Z_n = n\tau'(\theta), \quad n = 1, 2, \ldots,$$

where $\xi = \tau(\theta)$ denotes the rate of the process.

Infinitely divisible case

Let $\text{ED}^*(\theta, \lambda)$ denote an infinitely divisible additive family with $\Lambda = \mathbf{R}_+$. This case is associated with a continuous time stochastic process $\{Z(t) : t \geq 0\}$, which we now describe.

Define the increments of the process by

$$\Delta Z_i = Z(t_i) - Z(t_{i-1}),$$

where $0 = t_0 < \cdots < t_k$ are given time points. Let us assume that $Z(0) = 0$, and that for all k and all time points $t_1, \ldots, t_k$, the increments $\Delta Z_1, \ldots, \Delta Z_k$ are independent with distributions

$$\Delta Z_i \sim \text{ED}^*(\theta, \rho \Delta t_i), \quad i = 1, \ldots, k, \tag{3.21}$$

for some $\rho > 0$, where $\Delta t_i = t_i - t_{i-1}$. This process has stationary and independent increments, and we call it a **continuous time additive process**. It is an example of a type of process associated with infinitely divisible distributions, known as a **Lévy process**.

We now show that the above definition of the process is consistent. It follows, in turn, by Kolmogorov's Consistency Theorem (see e.g. Lamperti, 1966, pp. 14–15) that the process exists.

Proposition 3.8 *The continuous time additive process* $\{Z(t) : t \geq 0\}$ *is consistently defined by (3.21).*

Proof. Let $0 = t_0 < \cdots < t_k$ be given, and let the joint distribution of $Z(t_1), \ldots, Z(t_k)$ be defined by assuming that the increments $\Delta Z_1, \ldots, \Delta Z_k$ are independent with distribution defined by (3.21). We must show that the corresponding marginal distribution of any subset of $Z(t_1), \ldots, Z(t_k)$ is consistent with the definition of the process. It is sufficient to consider a subset of dimension $k - 1$, say the subset $Z(t_1), \ldots, Z(t_{k_0-1}), Z(t_{k_0+1}), \ldots, Z(t_k)$. By the independence of $\Delta Z_1, \ldots, \Delta Z_k$, we find that the new set of $k - 1$ increments

$$\Delta Z_1, \ldots, \Delta Z_{k_0} + \Delta Z_{k_0+1}, \ldots, \Delta Z_k$$

are independent. The additive form of the convolution formula gives

Table 3.1. *Three continuous exponential dispersion models*

	$N(\mu,\sigma^2)$	$\text{Ga}(\mu,\sigma^2)$	$\text{GHS}(\mu,\sigma^2)$
$c(y;\lambda)$	$\sqrt{\frac{\lambda}{2\pi}}e^{-\frac{\lambda y^2}{2}}$	$\frac{\lambda^\lambda y^{\lambda-1}}{\Gamma(\lambda)}$	$\frac{\lambda|\Gamma(\lambda(1+iy)/2)|^2}{\pi\Gamma(\lambda)2^{2-\lambda}}$
$\kappa(\theta)$	$\frac{1}{2}\theta^2$	$-\log(-\theta)$	$-\log(\cos\theta)$
$\tau(\theta)$	θ	$-1/\theta$	$\tan\theta$
$V(\mu)$	1	μ^2	$1+\mu^2$
Θ	$\mathbf{R}$	$\mathbf{R}_-$	$(-\pi/2,\pi/2)$
Ω	$\mathbf{R}$	$\mathbf{R}_+$	$\mathbf{R}$

$$\begin{aligned}\Delta Z_{k_0}+\Delta Z_{k_0+1} &\sim \text{ED}^*\{\theta,\rho(\Delta t_{k_0}+\Delta t_{k_0+1})\}\\ &= \text{ED}^*\{\theta,\rho(t_{k_0+1}-t_{k_0-1})\}.\end{aligned}$$

This proves the consistency of the definition. □

The marginal distribution of the process at time t is

$$Z(t) \sim \text{ED}^*(\theta,\rho t), \qquad (3.22)$$

with mean and variance

$$EZ(t)=\rho\tau(\theta)t \quad\text{and}\quad \text{var}\,Z(t)=\tau'(\theta)\rho t,$$

which are both linear in t. The rate of the process is hence $\xi = \rho\tau(\theta)$.

There also exist additive processes associated with additive models that are neither infinitely divisible, nor have $\Lambda=\mathbf{N}$, but we shall not consider such cases here.

3.3 Examples

We now consider five basic examples of exponential dispersion models, most of which we have met before, namely the normal, gamma, Poisson, binomial and negative binomial distributions. In particular, we study the form of the convolution formulas and additive processes. Together with the generalized hyperbolic secant (GHS) distribution (to be studied in Section 3.4.2), these distributions are known as the **Morris class**. Table 3.1 summarizes the three continuous distributions in the class, including the generalized hyperbolic secant distribution, and Table 3.2 summarizes the unit deviances for all six distributions in the Morris class.

Table 3.2. *Unit deviances for the six Morris models*

Model	Unit deviance
$N(\mu, \sigma^2)$	$(y-\mu)^2$
$Ga(\mu, \sigma^2)$	$2\left(\log\frac{\mu}{y} + \frac{y}{\mu} - 1\right)$
$Po(\mu)$	$2\left\{y\log\frac{y}{\mu} - (y-\mu)\right\}$
$Bi(m, \mu)$	$2\left\{y\log\frac{y}{\mu} + (1-y)\log\frac{1-y}{1-\mu}\right\}$
$Nb(p, \lambda)$	$2\left\{y\log\frac{y}{\mu} + (1+y)\log\frac{1+\mu}{1+y}\right\}$
$GHS(\mu, \sigma^2)$	$2y(\arctan y - \arctan \mu) + \log\frac{1+\mu^2}{1+y^2}$

3.3.1 Normal distribution

Let $N(\mu, \sigma^2)$ denote the normal distribution with mean μ and variance σ^2, and write the normal density function as

$$\begin{aligned}
p(y; \mu, \sigma^2) &= (2\pi\sigma^2)^{-\frac{1}{2}} \exp\left\{-\frac{1}{2\sigma^2}(y-\mu)^2\right\} \\
&= (2\pi\sigma^2)^{-\frac{1}{2}} \exp\left(-\frac{y^2}{2\sigma^2}\right) \exp\left\{\frac{1}{\sigma^2}\left(y\mu - \frac{1}{2}\mu^2\right)\right\}.
\end{aligned}$$

If we take $\theta = \mu$ and $\lambda = 1/\sigma^2$, we see that the normal distribution is a reproductive exponential dispersion model with unit cumulant function $\kappa(\theta) = \theta^2/2$. The corresponding mean value mapping is $\tau(\theta) = \theta$, and the unit variance function is $V(\mu) = 1$ defined on $\Omega = \mathbf{R}$.

The unit deviance of the normal distribution is the squared distance $(y-\mu)^2$. To understand the role of the unit deviance as a measure of fit, it may be useful to plot the unit deviance as a function of y for a given value of μ. For the normal distribution, the unit deviance has parabolic shape, whereas for other distributions the shape may be far from parabolic. Figure 3.1 shows such plots for the normal, gamma and generalized hyperbolic secant distributions. Such plots are useful for understanding the shape of the unit deviance, in particular asymmetry and the asymptotic behaviour of the unit deviance. The plots illustrate the asymmetry of the gamma and generalized hyperbolic unit deviances, the latter having a hyperbolic shape.

If $Y \sim N(\mu, \sigma^2)$ and we apply the duality transform $Z = \lambda Y$, we obtain the additive form of the normal distribution,

$$Z \sim N(\theta\lambda, \lambda) = \text{ED}^*(\theta, \lambda). \tag{3.23}$$

EXAMPLES

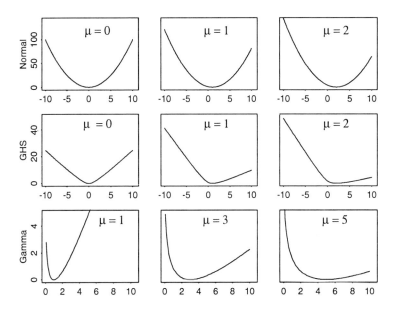

Figure 3.1. *Plots of the normal, GHS and gamma unit deviances.*

Note that we have used the scale transformation property of the normal distribution here.

Convolution

The additive version of the convolution formula for the normal distribution is

$$N(\theta\lambda_1, \lambda_1) + N(\theta\lambda_2, \lambda_2) = N\left\{\theta(\lambda_1 + \lambda_2), \lambda_1 + \lambda_2\right\}. \quad (3.24)$$

The general convolution formula for the normal distribution,

$$N(\mu_1, \sigma_1^2) + N(\mu_2, \sigma_2^2) = N\left(\mu_1 + \mu_2, \sigma_1^2 + \sigma_2^2\right), \quad (3.25)$$

is in fact more general than (3.24), but it is interesting to note that (3.25) may be derived from (3.24) by using the fact that μ is a location parameter.

The reproductive form of the convolution formula may be written as follows. If $Y_1, \ldots, Y_n$ are independent and $Y_i \sim N\left(\mu, \sigma^2/w_i\right)$, then

$$\frac{1}{w_+}\sum_{i=1}^{n} w_i Y_i \sim N\left(\mu, \frac{\sigma^2}{w_+}\right),$$

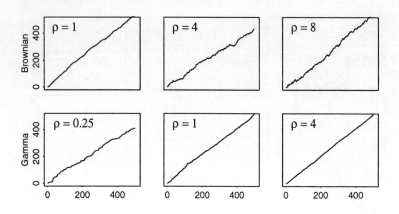

Figure 3.2. *Simulations of Brownian motion and the gamma process, all with rate $\xi = 1$.*

where $w_+ = w_1 + \cdots + w_n$.

Brownian motion

The additive process associated with the normal distribution is Brownian motion with drift. This may be seen by considering the additive form of the normal distribution, which shows that the process $Y(t)$, $t \geq 0$ is defined by $Y(0) = 0$, and assuming that the increments $\Delta Y_1, \ldots, \Delta Y_k$ are independent, with distribution

$$\Delta Y_i \sim N(\xi \Delta t_i, \rho \Delta t_i),$$

where $\xi = \theta \rho$ denotes the rate of the process. This corresponds to Brownian motion with drift ξ and variance parameter ρ. Figure 3.2 shows some simulations of Brownian motion.

3.3.2 Gamma distribution

From Chapter 1, we know that the gamma distribution $\text{Ga}(\mu, \sigma^2)$ is a reproductive exponential dispersion model with density for $y > 0$ given by

$$p(y; \mu, \sigma^2) = \frac{\lambda^\lambda}{\Gamma(\lambda)} y^{\lambda-1} \exp\left\{-\lambda \left(\frac{y}{\mu} + \log \mu\right)\right\},$$

where μ is the mean, $\sigma^2 = 1/\lambda$ is the squared coefficient of variation, and the canonical parameter is $\theta = -1/\mu$. The unit cumulant function is $\kappa(\theta) = -\log(-\theta)$. The model is infinitely divisible, and steep, with convex support $C = \Omega = \mathbf{R}_+$. The unit deviance is

$$d(y;\mu) = 2\left(\frac{y}{\mu} + \log\frac{\mu}{y} - 1\right),$$

and the unit variance function is $V(\mu) = \mu^2$, $\mu > 0$, giving the variance

$$\operatorname{var} Y = \sigma^2 \mu^2. \tag{3.26}$$

Figure 3.1 shows some plots of the gamma unit deviance.

A crucial property of the gamma distribution is the scale transformation property, which takes the following form for $c > 0$:

$$c\operatorname{Ga}(\mu, \sigma^2) = \operatorname{Ga}(c\mu, \sigma^2). \tag{3.27}$$

This result has several important consequences.

Using (3.27), we find that the normal approximation (3.11) takes the following form:

$$\operatorname{Ga}\left(\frac{\mu_0}{\sigma} + \mu, \sigma^2\right) - \frac{\mu_0}{\sigma} \xrightarrow{d} N\left(\mu, \mu_0^2\right) \quad \text{as} \quad \sigma^2 \to 0.$$

In a similar way to what we saw for the normal distribution, the scale transformation result implies that the gamma distribution is both an additive and a reproductive model. Thus, by (3.27), the duality transformation $Z = Y\lambda$ gives the following reproductive form of the gamma distribution:

$$Z \sim \operatorname{Ga}^*(\theta, \lambda) = \operatorname{Ga}\left(-\frac{\lambda}{\theta}, \frac{1}{\lambda}\right). \tag{3.28}$$

The corresponding additive form of the gamma density is

$$p^*(z; \theta, \lambda) = \frac{1}{\Gamma(\lambda)} z^{\lambda-1} \exp\left\{z\theta + \lambda \log(-\theta)\right\},$$

for $z > 0$, where $\theta < 0$. The mean and variance of Z are $EZ = -\lambda/\theta$ and $\operatorname{var} Z = \lambda/\theta^2$.

If $Y_1, \ldots, Y_n$ are independent and $Y_i \sim \operatorname{Ga}(\mu, \sigma^2/w_i)$, then the reproductive form of the convolution formula is

$$\frac{1}{w_+} \sum_{i=1}^n w_i Y_i \sim \operatorname{Ga}\left(\mu, \frac{\sigma^2}{w_+}\right), \tag{3.29}$$

where $w_+ = w_1 + \cdots + w_n$. The corresponding additive form of the convolution formula is as follows. Let $Z_1, \ldots, Z_n$ be independent and $Z_i \sim \operatorname{Ga}^*(\theta, \lambda)$, then

$$Z_1 + \cdots + Z_n \sim \operatorname{Ga}^*(\theta, \lambda_1 + \cdots + \lambda_n). \tag{3.30}$$

By (3.28), the two versions of the convolution formula are, of course, equivalent.

Exponential dispersion models that are closed under scale transformations are rather special; this topic will be explored further in Chapter 4.

Gamma process

The continuous time additive process for the gamma distribution is known as the **gamma process**. This process is defined by assuming that the increments $\Delta Z_1, \ldots, \Delta Z_k$ of the process $Z(t), t > 0$, are independent, with distribution

$$\Delta Z_i \sim \text{Ga}^*(\theta, \rho \Delta t_i),$$

where $\rho > 0$ and $Z(0) = 0$. The rate of the process is $\xi = -\rho/\theta$. Since the increments are always positive, this is a strictly increasing process. The marginal distribution of the process is $Z(t) \sim \text{Ga}^*(\theta, \rho t)$.

The gamma process is an example of a **pure jump process**, because in each time interval, the process has infinitely many jumps, most of which are very small, in spite of the fact that the increments follow a continuous distribution. Figure 3.2 shows some simulations of the gamma process, where the process appears to be piecewise continuous with an occasional jump. This is consistent with the pure jump nature of the process if we imagine that most of the jumps are so small as to make the process nearly continuous, except for a few major jumps. By comparison, Brownian motion is a continuous process, albeit nowhere differentiable.

3.3.3 Poisson distribution

We saw in Chapter 2 that the Poisson distribution is a natural exponential family. However, the result of Theorem 3.1 indicates that the Poisson distribution is also an additive exponential dispersion model, as we shall now see.

Consider a Poisson variable $Z \sim \text{Po}(\xi)$ with mean $\xi = \lambda e^\theta$, where $\lambda > 0$ and $\theta \in \mathbf{R}$. In this parametrization, the Poisson probability function has the form

$$p^*(z; \theta, \lambda) = \frac{\lambda^z}{z!} \exp\left(\theta z - \lambda e^\theta\right) \qquad (3.31)$$

for $z = 0, 1, \ldots$, which is an additive model $\text{ED}^*(\theta, \lambda)$ with unit cumulant function $\kappa(\theta) = e^\theta$. The parameters λ and θ are not

Table 3.3. *Three discrete exponential dispersion models*

	Po(λe^θ)	Bi(λ, μ)	Nb(p, λ)
$c^*(z; \lambda)$	$\frac{\lambda^z}{z!}$	$\binom{\lambda}{z}$	$\binom{\lambda+z-1}{z}$
$\kappa(\theta)$	e^θ	$\log\left(1 + e^\theta\right)$	$-\log\left(1 - e^\theta\right)$
$\tau(\theta)$	e^θ	$\frac{e^\theta}{1+e^\theta}$	$\frac{e^\theta}{1-e^\theta}$
$V(\mu)$	μ	$\mu(1-\mu)$	$\mu(1+\mu)$
Θ	**R**	**R**	**R**$_-$
Ω	**R**$_+$	$(0,1)$	**R**$_+$

identifiable in this case, but Theorem 3.1 shows that this is the only additive model where this happens. Table 3.3 summarizes the three discrete exponential dispersion models in the Morris class.

The unit variance function is $V(\mu) = \mu$, so that the natural exponential family ED$^*(\Theta, \lambda)$ has variance function

$$\xi \mapsto \lambda V(\xi/\lambda) = \lambda \xi/\lambda = \xi.$$

Hence, the variance function $\lambda V(\xi/\lambda)$ does not depend on λ, confirming that the families ED$^*(\Theta, \lambda)$ are all identical.

Suppose that $Z_1, \ldots, Z_n$ are independent and $Z_i \sim \text{Po}(\xi_i)$. By writing $\xi_i = \lambda_i e^\theta$ for some fixed θ, we may apply the additive form of the convolution formula, obtaining

$$\begin{aligned} Z_1 + \cdots + Z_n &\sim \text{Po}\left\{(\lambda_1 + \cdots + \lambda_n)e^\theta\right\} \\ &= \text{Po}(\xi_1 + \cdots + \xi_n). \end{aligned}$$

Hence, the standard convolution formula for the Poisson distribution is a special case of the convolution formula for exponential dispersion models.

The Poisson unit deviance is

$$d(y; \mu) = 2\left\{y \log \frac{y}{\mu} - (y - \mu)\right\};$$

some plots of it are shown in Figure 3.3.

The continuous time additive process $Z(t)$ corresponding to the Poisson distribution is defined by the following distribution of the increments:

$$\Delta Z_i \sim \text{Po}\left(\rho e^\theta \Delta t_i\right),$$

starting from $Z(0) = 0$. The corresponding marginal distribution is $Z(t) \sim \text{Po}\left(\rho e^\theta t\right)$. This is the Poisson process with rate $\xi = \rho e^\theta$. Figure 3.4 shows some simulations of the Poisson process.

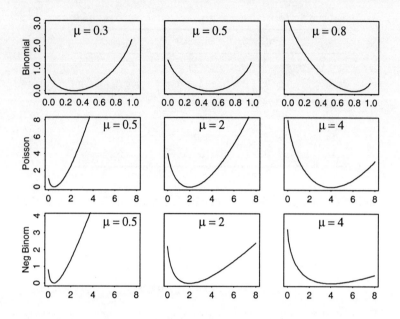

Figure 3.3. *Plots of the Poisson, binomial and negative binomial unit deviances.*

Convergence results

Let $Z \sim \text{Po}(\xi)$ be a Poisson random variable, and let $\xi = \lambda e^\theta$. Then letting λ tend to infinity for θ fixed amounts to letting ξ tend to infinity. Hence, the convergence formula (3.12) takes the following form for the Poisson distribution:

$$\frac{Z - \xi}{\sqrt{\xi}} \xrightarrow{d} N(0, 1) \quad \text{as} \quad \xi \to \infty.$$

This is the usual normal approximation to the Poisson distribution. A large value of the index parameter λ is hence equivalent to a large mean, and the normal approximation holds for large values of the mean.

3.3.4 Binomial distribution

Let $Z \sim \text{Bi}(m, \mu)$ denote a binomial random variable giving the number of successes in m independent Bernoulli trials with probability parameter μ. Writing the probability function of Z in terms

EXAMPLES 93

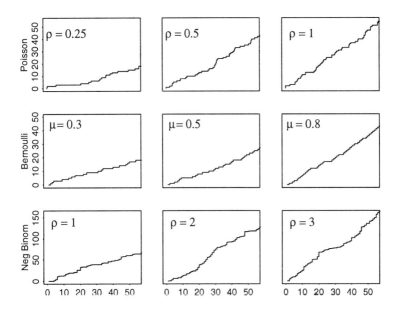

Figure 3.4. *Simulations of the Poisson, Bernoulli and negative binomial processes. The Poisson and negative binomial processes have rate ρ and the binomial processes have rate μ.*

of the logistic parameter $\theta = \log \mu/(1-\mu)$ we obtain

$$\begin{aligned} p^*(z; m, \mu) &= \binom{m}{z} \mu^z (1-\mu)^{m-z} \\ &= \binom{m}{z} \exp\left\{\theta z - m \log(1 + e^\theta)\right\}, \end{aligned}$$

for $z = 0, \ldots, m$. This shows that the binomial distribution is an additive exponential dispersion model $\mathrm{ED}^*(\theta, m)$ with canonical parameter $\theta \in \mathbf{R}$ and index parameter $m \in \mathbf{N}$. Figure 3.3 shows some plots of the binomial unit deviance

$$d(y; \mu) = 2 \left\{ y \log \frac{y}{\mu} + (1-y) \log \frac{1-y}{1-\mu} \right\}.$$

One may show that the index set for the binomial additive model is $\Lambda = \mathbf{N}$, in other words that non-integer values of m are impossible. The binomial distribution is not infinitely divisible (see Exercise 3.8), but the proof that $\Lambda = \mathbf{N}$ is delicate. An argument based on a result due to Jørgensen (1986) was given in the reply to the discussion of Jørgensen (1987a).

The unit cumulant function is $\kappa(\theta) = \log(1 + e^\theta)$, which gives $\tau(\theta) = e^\theta/(1 + e^\theta)$ and $\Omega = (0, 1)$. The convex support of Z is $C_m = [0, m]$, giving an example of a steep additive model where the convex support depends on the index parameter m.

From $\tau'(\theta) = e^\theta/(1 + e^\theta)^2$ it follows that the unit variance function is
$$V(\mu) = \mu(1 - \mu), \quad \mu \in (0, 1).$$
If we express the variance in terms of the mean $\xi = m\mu$, we obtain the expression
$$\text{var}\, Z = mV(\xi/m) = \xi(1 - \xi/m).$$

If $Z_1, \ldots, Z_n$ are independent and $Z_i \sim \text{Bi}(m_i, \mu)$ for $i = 1, \ldots, n$, the convolution formula (3.14) gives the standard convolution formula for the binomial distribution,
$$Z_1 + \cdots + Z_n \sim \text{Bi}(m_1 + \cdots + m_n, \mu).$$

The discrete time additive process corresponding to the binomial distribution is the Bernoulli process Z_n, corresponding to the cumulated number of successes out of n trials for a sequence of independent Bernoulli trials. The process starts from $Z_0 = 0$, the increments having a Bernoulli distribution
$$Z_n - Z_{n-1} \sim \text{Bi}(1, \mu).$$
Figure 3.4 shows some simulations of the Bernoulli process.

For $Z \sim \text{Bi}(m, \mu)$, the normal approximation formula (3.12) gives
$$\frac{Z - m\mu}{\sqrt{m}} \xrightarrow{d} N\{0, \mu(1 - \mu)\} \quad \text{as} \quad m \to \infty,$$
which give the usual normal approximation to the binomial distribution (de Moivre-Laplace's formula).

From Theorem 3.6 we find that
$$\text{Bi}(m, \xi/m) \xrightarrow{d} \text{Po}(\xi) \quad \text{as} \quad m \to \infty,$$
which is the standard Poisson approximation to the binomial distribution.

3.3.5 Negative binomial distribution

Let Z be a negative binomial random variable, with probability parameter p and shape parameter λ. The probability function of

EXAMPLES

Z is, for $z = 0, 1, \ldots$,

$$\begin{aligned} p^*(z; p, \lambda) &= \binom{\lambda + z - 1}{z} p^z (1-p)^\lambda \\ &= \binom{\lambda + z - 1}{z} \exp\{\theta z + \lambda \log(1 - e^\theta)\}, \end{aligned}$$

which is an additive model with canonical parameter $\theta = \log p < 0$ and index parameter $\lambda > 0$. When parametrized by λ and p, the negative binomial distribution is denoted by $\mathrm{Nb}(p, \lambda)$. For λ integer, Z is the number of successes in a sequence of Bernoulli trials before the λth failure, p being the probability of success.

The negative binomial distribution has unit cumulant function

$$\kappa(\theta) = -\log(1 - e^\theta),$$

showing a clear duality with the binomial model, as highlighted in Table 3.3. The mean value mapping is

$$\tau(\theta) = \frac{e^\theta}{1 - e^\theta} = \frac{p}{1-p},$$

giving the mean domain $\Omega = \mathbf{R}_+$ and mean

$$\xi = EZ = \frac{\lambda p}{1-p}.$$

The model is hence steep, with convex support $[0, \infty)$.

By differentiating τ we obtain

$$\tau'(\theta) = \frac{e^\theta}{(1 - e^\theta)^2} = \frac{p}{(1-p)^2},$$

giving the unit variance function

$$V(\mu) = \mu(1 + \mu),$$

for $\mu > 0$. The variance expressed in terms of the mean ξ is

$$\operatorname{var} Z = \lambda V\left(\frac{\xi}{\lambda}\right) = \xi\left(1 + \frac{\xi}{\lambda}\right),$$

and in terms of the probability parameter,

$$\operatorname{var} Z = \frac{\lambda p}{(1-p)^2}.$$

The unit deviance, shown in Figure 3.3, is defined by

$$d(y; \mu) = 2\left\{y \log \frac{y(1+\mu)}{\mu(1+y)} + \log \frac{1+\mu}{1+y}\right\}, \qquad (3.32)$$

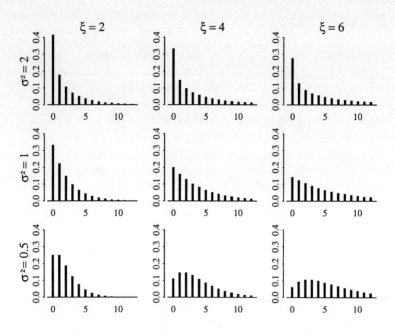

Figure 3.5. *Some negative binomial probability functions.*

for $y, \mu > 0$. Some plots of the negative binomial probability function are shown in Figure 3.5.

Convergence results

The negative binomial distribution exhibits three different types of asymptotic behaviour. The first is asymptotic normality. If $Z \sim \text{Nb}(p, \lambda)$, we obtain, from (3.12), convergence to normality for λ large and p fixed, which we may write as follows:

$$\frac{1}{\sqrt{\lambda}} \left(Z - \frac{\lambda p}{1-p} \right) \xrightarrow{d} N\left\{ 0, \frac{p}{(1-p)^2} \right\} \quad \text{as} \quad \lambda \to \infty.$$

The second result says that a negative binomial distribution with mean ξ is asymptotically Poisson for large λ, which we may write as follows:

$$\text{Nb}\left(\frac{\xi}{\lambda + \xi}, \lambda \right) \xrightarrow{d} \text{Po}(\xi) \quad \text{as} \quad \lambda \to \infty.$$

QUADRATIC VARIANCE FUNCTIONS 97

The third result shows convergence to the gamma distribution,

$$\frac{1}{c}\text{Nb}\left(\frac{c\mu}{1+c\mu},\lambda\right) \xrightarrow{d} \text{Ga}\left(\mu,\frac{1}{\lambda}\right) \quad \text{as} \quad c \to \infty.$$

This result may be interpreted as saying that the negative binomial distribution with large mean ξ and index parameter λ may be approximated by a $\text{Ga}(\xi, \lambda^{-1})$ distribution. It may be proved using Theorem 3.5, see Exercise 3.22.

Convolution and the additive process

The convolution formula (3.14) takes the following form for the negative binomial distribution. If $Z_1, \ldots, Z_n$ are independent, and $Z_i \sim \text{Nb}(p, \lambda_i)$ for $i = 1, \ldots, n$, then

$$Z_1 + \cdots + Z_n \sim \text{Nb}(p, \lambda_1 + \cdots + \lambda_n),$$

which is the standard convolution formula for the negative binomial distribution.

The continuous time additive process corresponding to the negative binomial distribution, called the **negative binomial process**, is defined as the process $Z(t)$ starting in $Z(0) = 0$ with increments distributed as

$$\Delta Z_i \sim \text{Nb}(p, \rho \Delta t_i),$$

where $\rho > 0$. The rate of the process is $\xi = \rho/\left(p^{-1}-1\right)$. The embedded discrete time process obtained for $\rho = 1$ and $t \in \mathbf{N}$ is the geometric process, which may be interpreted as the cumulative number of successes until the tth failure in a sequence of independent Bernoulli trials. Figure 3.4 shows some simulations of the negative binomial process.

3.4 Quadratic variance functions

By a quadratic variance function we mean a variance function

$$V(\mu) = a\mu^2 + b\mu + c, \quad \mu \in \Omega, \tag{3.33}$$

which is a polynomial of degree at most two. All five examples of exponential dispersion models considered in Section 3.3 have quadratic variance functions, see Tables 3.1 and 3.3.

A sixth exponential dispersion model with quadratic variance function is the generalized hyperbolic secant distribution, which we describe in Section 3.4.2. We now show that up to affine transformations, these six distributions are the only exponential dispersion models with quadratic variance functions, a result due to Morris (1982).

3.4.1 Morris classification

To state Morris' result, we consider two types of transformations of a variance function. If V with domain Ω is a given variance function for a natural exponential family $\mathcal{P}$, then the affine transformation $y \mapsto \alpha + \beta y$ ($\beta \neq 0$) transforms the family into a new natural exponential family with variance function $\beta^2 V\{(\mu - \alpha)/\beta\}$, having domain $\alpha + \beta \Omega$ (see Exercise 3.1).

The second type of transformation comes from generating a reproductive exponential dispersion model from $\mathcal{P}$, which multiplies V by a constant $\sigma^2 \in \Lambda^{-1}$. By combining these two transformations, we obtain the variance function

$$\gamma V \left(\frac{\mu - \alpha}{\beta} \right), \quad (3.34)$$

defined on $\alpha + \beta \Omega$, where $\gamma = \sigma^2 \beta^2$. In the case $\Lambda = \mathbf{R}_+$, any value of $\gamma > 0$ may be obtained for given α and β, but only certain values of γ may be obtained for given β if the distribution is not infinitely divisible. We refer to the two transformations just described as **affine transformations** and **division/convolution**, respectively. The latter terminology comes from the proof of Theorem 3.7.

Morris' classification involves the following six unit variance functions:

1. $V(\mu) = 1 \quad (\Omega = \mathbf{R})$;
2. $V(\mu) = \mu \quad (\Omega = \mathbf{R}_+)$;
3. $V(\mu) = \mu(1 - \mu) \quad (\Omega = (0,1))$;
4. $V(\mu) = \mu(1 + \mu) \quad (\Omega = \mathbf{R}_+)$;
5. $V(\mu) = \mu^2 \quad (\Omega = \mathbf{R}_+)$;
6. $V(\mu) = 1 + \mu^2 \quad (\Omega = \mathbf{R})$.

The six cases correspond to the normal distribution with variance 1, the Poisson distribution, the Bernoulli distribution, the geometric (unit negative binomial) distribution, the exponential distribution and the hyperbolic secant distribution, respectively.

Theorem 3.9 (Morris) *Any quadratic variance function may be obtained by affine transformation and division/convolution from the above six unit variance functions.*

QUADRATIC VARIANCE FUNCTIONS

Proof. The proof consists in describing the set of models obtained by applying the transformation (3.34) to each of the above six cases.

1. Applying the transformation to $V(\mu) = 1$ gives the constant variance function γ for any $\gamma > 0$. This accounts for all constant variance functions.
2. Applying the transformation to $V(\mu) = \mu$ gives the variance function $V(\mu) = \gamma(\mu-\alpha)/\beta$. This accounts for all affine variance functions that are not constant.
3. In the Bernoulli case $V(\mu) = \mu(1-\mu)$, we must take into account that the distribution is not infinitely divisible. First, by division/convolution, we obtain the binomial distribution. By the duality transformation, we may consider its reproductive form, with variance function $\mu(1-\mu)/m$ where $m \in \mathbf{N}$. After an affine transformation, this gives the variance function
$$V(\mu) = \frac{1}{m}(\mu - \alpha)(\alpha + \beta - \mu),$$
which is a quadratic polynomial with two distinct roots and negative leading coefficient $a = -1/m$. By choosing α and β appropriately, we may obtain any two roots. However, the leading coefficient is restricted to the choice $-1/a \in \mathbf{N}$. But no other values of a are possible, because the binomial index set is $\Lambda = \mathbf{N}$.
4. Applying the transformation (3.34) to $V(\mu) = \mu(1+\mu)$ $(\mu > 0)$ gives the variance function
$$V(\mu) = \frac{\gamma}{\beta^2}(\mu - \alpha)(\mu - \alpha + \beta).$$
This accounts for all quadratic polynomials with two distinct roots and positive leading coefficient a.
5. Applying the transformation to $V(\mu) = \mu^2$ $(\mu > 0)$ gives the variance function
$$V(\mu) = \frac{\gamma}{\beta^2}(\mu - \alpha)^2.$$
This accounts for all quadratic polynomials with a double root.
6. Applying the transformation to $V(\mu) = 1 + \mu^2$, we obtain the variance function
$$V(\mu) = \gamma + \frac{\gamma}{\beta^2}(\mu - \alpha)^2, \quad \mu \in \mathbf{R}, \qquad (3.35)$$
with $\gamma > 0$. This accounts for all quadratic variance functions with no real roots.

We have now considered all possible polynomials of degree at most two. Taking into account that the mean domain must be an interval, we have found all possible quadratic variance functions, with one possible exception. Could we obtain a new natural exponential family from a given variance function by restricting its mean domain? The answer is no, because, by the proof of the uniqueness theorem, we would obtain a subset of the given family, and two natural exponential families with an element in common are identical. This concludes the proof. □

Note that by division/convolution, we obtain six exponential dispersion models from the six unit variance functions, namely the five distributions described in Section 3.3 and the generalized hyperbolic secant distribution. Note also that the duality transformation is affine; hence we may choose between the additive and reproductive forms of each model, to conform with the forms chosen in Sections 3.3.4 and 3.3.5. Hence Morris' result says that up to affine transformations, these six exponential dispersion models are the only ones with quadratic variance functions.

3.4.2 Generalized hyperbolic secant distribution

In order to complete Morris' classification, we need to show that the unit variance function $V(\mu) = 1 + \mu^2$ corresponds to an exponential dispersion model, and that this model is infinitely divisible. In doing so, we take the opportunity to show how a natural exponential family may be recovered from its variance function. We hence follow the steps of the proof of the uniqueness theorem for variance functions in Chapter 2.

The first step is to solve the differential equation

$$\frac{\partial \tau^{-1}}{\partial \mu} = \frac{1}{1 + \mu^2}, \quad \mu \in \mathbf{R},$$

which has solution

$$\tau(\theta) = \tan \theta, \quad |\theta| < \pi/2, \tag{3.36}$$

where we ignore the arbitrary constant, which we know does not affect the result. Next, we solve the differential equation

$$\kappa'(\theta) = \tan \theta, \quad |\theta| < \pi/2,$$

which, again ignoring the arbitrary constant, gives

$$\kappa(\theta) = -\log\left(\cos \theta\right), \quad |\theta| < \pi/2.$$

QUADRATIC VARIANCE FUNCTIONS 101

Note that (3.36) implies $\Omega = \mathbf{R}$; hence the convex support is $C = \mathbf{R}$ and the family is steep. The canonical parameter domain is $\Theta = (-\pi/2, \pi/2)$.

The natural exponential family corresponding to this cumulant function is known as the **hyperbolic secant distribution**. It is continuous and has density

$$p(y;\theta) = \frac{\exp\{\theta y + \log(\cos\theta)\}}{2\cosh(\pi y/2)}, \quad y \in \mathbf{R}. \tag{3.37}$$

This distribution may be derived from the beta distribution with parameters $1/2 + \theta\pi$ and $1/2 - \theta/\pi$, which has density

$$f(u;\theta) = \frac{\cos\theta}{\pi} u^{-\frac{1}{2}+\frac{\theta}{\pi}}(1-u)^{-\frac{1}{2}-\frac{\theta}{\pi}}, \tag{3.38}$$

where $0 < u < 1$. By the transformation $y = \pi^{-1}\log\{u/(1-u)\}$, (3.38) turns into the density (3.37) (Exercise 3.21). The particular form of the beta function in formula (3.38) comes from

$$\frac{\Gamma(1)}{\Gamma\left(\frac{1}{2}-\frac{\theta}{\pi}\right)\Gamma\left(\frac{1}{2}+\frac{\theta}{\pi}\right)} = \frac{\sin\left(\theta+\frac{\pi}{2}\right)}{\pi}$$

$$= \frac{\cos\theta}{\pi},$$

which may be shown by the reflection formula for the gamma function (Abramowitz and Stegun, 1972, p. 256).

The hyperbolic secant distribution is infinitely divisible (Feller, 1971, p. 567); hence it generates an additive exponential dispersion model with index set $\Lambda = \mathbf{R}_+$, called the **generalized hyperbolic secant distribution,** with density

$$p^*(z;\theta,\lambda) = c^*(z;\lambda)\exp\{\theta z + \lambda\log(\cos\theta)\} \tag{3.39}$$

for $z \in \mathbf{R}$. Here $c^*(z;\lambda)$ is a symmetric density function (Harkness and Harkness, 1968) given for $z \in \mathbf{R}$ by

$$\begin{aligned} c^*(z;\lambda) &= \frac{2^{\lambda-2}\left|\Gamma(\lambda/2+iz/2)\right|^2}{\pi\Gamma(\lambda)} \\ &= \frac{2^{\lambda-2}\Gamma^2(\lambda/2)}{\pi\Gamma(\lambda)} \prod_{j=0}^{\infty}\left\{1+\left(\frac{z}{\lambda+2j}\right)^2\right\}^{-1}. \end{aligned} \tag{3.40}$$

Formula (3.40) follows from a formula for the gamma function of complex argument (Abramowitz and Stegun, 1972, p. 256), correcting an error in the corresponding formula of Morris (1982, p. 73).

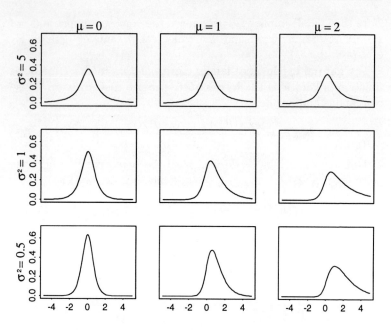

Figure 3.6. *Some generalized hyperbolic secant density functions.*

Let GHS(μ, σ^2) denote the reproductive version of the generalized hyperbolic secant model, where $\mu = \tan\theta$ is the mean and $\sigma^2 = 1/\lambda$ the dispersion parameter. The density function is

$$p(y; \mu, \sigma^2) = \lambda c^*(\lambda y; \lambda) \exp\left[\lambda \left\{ y \arctan \mu - \frac{1}{2}\log\left(1 + \mu^2\right) \right\}\right]$$

for $y \in \mathbf{R}$. The unit deviance has the following form:

$$d(y; \mu) = 2y\left(\arctan y - \arctan \mu\right) + \log \frac{1 + \mu^2}{1 + y^2}.$$

Figure 3.6 shows some plots of generalized hyperbolic secant densities. The plots show that μ, besides being the mean, controls the skewness of the distribution. Figure 3.1 shows plots of the unit deviance. Morris (1982) gives more details about the generalized hyperbolic secant density, and in Exercise 3.23 we calculate its coefficients of skewness and kurtosis.

By Theorem 3.5 the generalized hyperbolic secant distribution satisfies the convergence formula

$$c^{-1}\text{GHS}(c\mu, \sigma^2) \xrightarrow{d} \text{Ga}(\mu, \sigma^2) \quad \text{as } c \to \infty. \tag{3.41}$$

This result may be interpreted as saying that the $\text{GHS}(\mu, \sigma^2)$ distribution behaves asymptotically as the gamma distribution $\text{Ga}(\mu, \sigma^2)$ when μ is large.

3.5 Saddlepoint approximation

We now consider the saddlepoint approximation for exponential dispersion models, which is an important tool for asymptotic theory. We discuss the discrete and continuous cases separately.

3.5.1 Continuous case

Let us consider a continuous reproductive exponential dispersion model $Y \sim \text{ED}(\mu, \sigma^2)$ with probability density function in standard form, for $Y \in C$ and $\mu \in \Omega$,

$$p(y; \mu, \sigma^2) = a(y; \sigma^2) \exp\left\{-\frac{1}{2\sigma^2} d(y; \mu)\right\}, \tag{3.42}$$

and let $\lambda = 1/\sigma^2$ denote the index parameter, as usual. Let us write the unit deviance as follows, for $y \in \Omega$ and $\tilde{\theta} = \tau^{-1}(y) \in \text{int } \Theta$,

$$d(y; \mu) = 2\left[y\tilde{\theta} - \kappa(\tilde{\theta}) - y\tau^{-1}(\mu) + \kappa\left\{\tau^{-1}(\mu)\right\}\right]. \tag{3.43}$$

The saddlepoint approximation then takes the following form, for $y \in \Omega$:

$$p(y; \mu, \sigma^2) = \{2\pi\sigma^2 V(y)\}^{-\frac{1}{2}} \exp\left\{-\frac{1}{2\sigma^2} d(y; \mu)\right\} \{1 + O(\sigma^2)\} \tag{3.44}$$

as $\sigma^2 \to 0$. The proof of the saddlepoint approximation is sketched below.

Due to the factor $V^{-1/2}(y)$ the saddlepoint approximation is not defined outside Ω. However, because the variance of Y is $\sigma^2 V(\mu)$, the probability mass outside Ω is small when σ^2 is small. Furthermore, in the steep case we have $C = \Omega$ (because the model is continuous), so in the continuous steep case there is no probability mass outside Ω.

The original derivation of the saddlepoint approximation by Daniels (1954) was as an approximation for the density of the sample mean for a given probability density function. Daniels did not mention the exponential family, although it is implicit in his derivation. Note that any exponential dispersion model with large λ may be thought of as the distribution of a sample average for a large sample, due to the reproductive version of the convolution formula.

Recall that the saddlepoint approximation is called uniform on compacts if the convergence in

$$\sigma a(y; \sigma^2) \to \{2\pi V(y)\}^{-\frac{1}{2}} \quad \text{as} \quad \sigma^2 \to 0 \tag{3.45}$$

is uniform in y on compact subsets of Ω. The following result is due to Barndorff-Nielsen and Cox (1979).

Theorem 3.10 (Saddlepoint approximation) *Suppose there exists a λ_0 in Λ such that for all $\lambda > \lambda_0$ the density (3.42) is bounded. Then the saddlepoint approximation is uniform on compacts.*

In some cases, the saddlepoint approximation may be uniform in y on all of Ω, as Daniels (1954) showed. Several authors have studied this question recently, see Jensen (1988), Routledge and Tsao (1995) and references therein.

As the above results show, the form of the saddlepoint approximation is very simple when the density is expressed in standard form. Suppose that we instead have a density of the form

$$p(y; \mu, \sigma^2) = c(y; \lambda) \exp\left[\lambda \left\{y\theta - \kappa(\theta)\right\}\right].$$

Then the saddlepoint approximation amounts to the following:

$$c(y; \lambda) \exp\left[\lambda \left\{y\tilde{\theta} - \kappa(\tilde{\theta})\right\}\right] \sim [\lambda/\{2\pi V(y)\}]^{\frac{1}{2}}$$

for λ large. What makes the saddlepoint approximation especially useful is that the 'ugly' part of the density is often hidden in $c(y; \lambda)$, which the approximation neatly gets rid of.

Example 3.1 Let us consider the saddlepoint approximation for the gamma distribution. The unit deviance is

$$d(y; \mu) = 2(\log \mu/y + y/\mu - 1),$$

and the dispersion model form of the density is

$$a(y; \sigma^2) \exp\left\{-\frac{1}{2\sigma^2} d(y; \mu)\right\},$$

where

$$a(y; \lambda^{-1}) = \frac{\lambda^\lambda e^{-\lambda}}{\Gamma(\lambda)} y^{-1}.$$

SADDLEPOINT APPROXIMATION

The saddlepoint approximation is hence equivalent to Stirling's formula,
$$\Gamma(\lambda) \sim \sqrt{2\pi} \lambda^{\lambda-1/2} e^{-\lambda}. \tag{3.46}$$
Since the saddlepoint approximation is proportional to the density function, the renormalized saddlepoint approximation is exact. The ordinary saddlepoint approximation is quite accurate for the gamma distribution, even for moderately small values of λ. For example, the relative error of (3.46) is about 1% for $\lambda = 8$. □

Proof of continuous case

We now sketch a proof of the saddlepoint approximation in the continuous case.

The reproductive model $Y \sim \mathrm{ED}(\mu, \sigma^2)$ has characteristic function
$$\varphi(t; \theta, \lambda) = \exp[\lambda\{\kappa(\theta + it/\lambda) - \kappa(\theta)\}].$$
If $\varphi(t; \theta, \lambda)$ is absolutely integrable then, by the Fourier inversion theorem, the probability density function of Y is
$$p(y; \mu, \sigma^2) = \frac{1}{2\pi} \int_{-\infty}^{\infty} \varphi(t; \theta, \lambda) e^{-ity} \, dt,$$
where i is the complex imaginary unit. Hence the probability density function of Y becomes
$$\begin{aligned} p(y; \mu, \sigma^2) &= \frac{1}{2\pi} \int_{-\infty}^{\infty} \exp[\lambda\{\kappa(\theta + it/\lambda) - \kappa(\theta)\} - ity] \, dt \\ &= \frac{\lambda}{2\pi} \int_{-\infty}^{\infty} \exp[\lambda\{\kappa(\theta + is) - \kappa(\theta) - isy\}] \, ds, \end{aligned}$$
after making the substitution $s = t/\lambda$. We now choose μ, y in Ω, such that both $\theta = \tau^{-1}(\mu)$ and $\tilde{\theta} = \tau^{-1}(y)$ belong to int Θ. Since the integrand is analytic, we may move the path of integration from $(-\infty, \infty)$ to $i(\theta - \tilde{\theta}) + (-\infty, \infty)$. The density then becomes
$$\frac{\lambda}{2\pi} \int_{-\infty}^{\infty} \exp\left[\lambda\{\kappa(\tilde{\theta} + is) - (\tilde{\theta} + is)y + \theta y - \kappa(\theta)\}\right] ds \tag{3.47}$$
$$= \frac{\lambda}{2\pi} \exp\left[\lambda\{\theta y - \kappa(\theta)\}\right] \int_{-\infty}^{\infty} \exp\left[\lambda\{\kappa(\tilde{\theta} + is) - (\tilde{\theta} + is)y\}\right] ds.$$
Note that the last integral gives the following expression for $c(y; \lambda)$:
$$c(y; \lambda) = \frac{\lambda}{2\pi} \int_{-\infty}^{\infty} \exp\left[\lambda\{\kappa(\tilde{\theta} + is) - (\tilde{\theta} + is)y\}\right] ds.$$

We now consider the term in curly brackets in (3.47). By introducing the unit deviance and in turn expanding κ around $\tilde{\theta}$, this term becomes

$$\kappa(\tilde{\theta} + is) - (\tilde{\theta} + is)y + \tilde{\theta}y - \kappa(\tilde{\theta}) - \frac{1}{2}d(y; \mu)$$

$$\simeq \frac{1}{2}(is)^2 \kappa''(\tilde{\theta}) - \frac{1}{2}d(y; \mu)$$

$$= -\frac{1}{2}s^2 V(y) - \frac{1}{2}d(y; \mu).$$

From the result

$$\int_{-\infty}^{\infty} \exp\left\{-\frac{\lambda}{2}V(y)s^2\right\} ds = \sqrt{\frac{2\pi}{\lambda V(y)}},$$

and noting that the factor involving the unit deviance is constant as a function of s, we obtain the approximation

$$p(y; \mu, \sigma^2) \sim \frac{\lambda}{2\pi} \exp\left\{-\frac{\lambda}{2}d(y; \mu)\right\} \sqrt{\frac{2\pi}{\lambda V(y)}}$$

$$= \left\{\frac{\lambda}{2\pi V(y)}\right\}^{1/2} \exp\left\{-\frac{\lambda}{2}d(y; \mu)\right\},$$

which is the main term of the saddlepoint approximation (3.44).

The terminology 'saddlepoint approximation' comes from the fact that the exponent of the integrand has a saddlepoint at $\tilde{\theta}$.

3.5.2 Discrete case

Consider an integer-valued additive model $Z \sim \mathrm{ED}^*(\theta, \lambda)$ with probability function for $z \in \mathbf{Z}$ given by

$$\begin{aligned} p^*(z; \xi, \sigma^2) &= c^*(z; \lambda) \exp\{\theta z - \lambda \kappa(\theta)\} \\ &= a^*(z; \sigma^2) \exp\left\{-\frac{1}{2\sigma^2} d(z\sigma^2; \xi\sigma^2)\right\}. \end{aligned}$$

Here $\xi = \lambda \tau(\theta)$ denotes the mean of Z. The simplest way to arrive at the saddlepoint approximation in this case is to apply the duality transform $Z = Y/\sigma^2$ to (3.44). This gives the following form for $p^*(z; \xi, \sigma^2)$:

$$\frac{\sigma}{\sqrt{2\pi V(z\sigma^2)}} \exp\left\{-\frac{1}{2\sigma^2} d(z\sigma^2; \xi\sigma^2)\right\} \{1 + O(\sigma^2)\} \quad (3.48)$$

SADDLEPOINT APPROXIMATION

$$= \frac{1}{\sqrt{2\pi\lambda V(z/\lambda)}} \exp\left\{\theta z - \lambda\kappa(\theta) - \tilde{\theta}z + \kappa(\tilde{\theta})\right\}\left\{1 + O(\sigma^2)\right\},$$

for $z \in \mathbf{Z} \cap \sigma^{-2}\Omega$, where $\tilde{\theta} = \tau^{-1}(z\sigma^2)$. In this case the saddlepoint approximation is called uniform on compacts if the convergence in

$$\sigma^{-1}a^*(y\sigma^{-2};\sigma^2) \to \{2\pi V(y)\}^{-\frac{1}{2}} \quad \text{as} \quad \sigma^2 \to 0$$

is uniform in y on compact subsets of Ω. The following result is due to Barndorff-Nielsen and Cox (1979).

Theorem 3.11 (Saddlepoint approximation) *Suppose that the support of Z is not contained in any sublattice of $\mathbf{Z}$. Then the saddlepoint approximation is uniform on compacts.*

The proof of the saddlepoint approximation is sketched below.

In the special case where the support of the distribution is $\mathbf{N}_0$, special care is needed for the value $z = 0$, where the variance function is zero, and the saddlepoint approximation hence infinite. McCullagh and Nelder (1989) suggested the following ad hoc modification of (3.48):

$$p^*(z;\xi,\sigma^2) \sim \frac{\sigma}{\sqrt{2\pi V\{(z+c)\sigma^2\}}} \exp\left\{-\frac{1}{2\sigma^2}d(z\sigma^2;\xi\sigma^2)\right\}, \tag{3.49}$$

where c is a small positive constant, which they chose to be $1/3$.

Proof of discrete case

In the discrete integer-valued case the inversion formula for a characteristic function φ takes the form

$$p(z) = \frac{1}{2\pi}\int_{-\pi}^{\pi}\varphi(t)e^{-itz}dt,$$

for $z \in \mathbf{Z}$. For an integer-valued additive exponential dispersion model we hence obtain

$$p^*(z;\xi,\sigma^2) = \frac{1}{2\pi}\int_{-\pi}^{\pi}\exp[\lambda\{\kappa(\theta + it) - \kappa(\theta)\} - itz]\,dt.$$

By the analyticity and periodicity of the integrand, proceeding by analogy with the continuous case, we may change the path of integration to $(-\pi,\pi) + i(\theta - \tilde{\theta})$. Hence, for $z/\lambda = \tau(\tilde{\theta}) \in \Omega$, the probability function $p(z;\theta,\lambda)$ may be approximated as follows:

$$\frac{1}{2\pi}\int_{-\pi}^{\pi}\exp\left[\lambda\{\kappa(\tilde{\theta}+it) - \kappa(\theta)\} - (\tilde{\theta}+it)z + \theta z\right]dt$$

$$\sim \frac{1}{2\pi} \exp\left\{-\frac{\lambda}{2} d\left(\frac{z}{\lambda}; \frac{\xi}{\lambda}\right)\right\} \int_{-\pi}^{\pi} \exp\left\{-\frac{\lambda}{2} V\left(\frac{z}{\lambda}\right) t^2\right\} dt$$

$$\sim \frac{1}{2\pi} \exp\left\{-\frac{\lambda}{2} d\left(\frac{z}{\lambda}; \frac{\xi}{\lambda}\right)\right\} \int_{-\infty}^{\infty} \exp\left\{-\frac{\lambda}{2} V\left(\frac{z}{\lambda}\right) t^2\right\} dt$$

$$= \{2\pi \lambda V(z/\lambda)\}^{-\frac{1}{2}} \exp\left\{-\frac{\lambda}{2} d\left(\frac{z}{\lambda}; \frac{\xi}{\lambda}\right)\right\},$$

for λ tending to infinity. This gives the leading term of (3.48).

3.6 Residuals and tail area approximations

3.6.1 Residuals

We now consider saddlepoint approximations for the distribution function of a dispersion model. This subject is closely related to the study of residuals and their asymptotic normality. The results have important applications for model checking in generalized linear models, although we shall not pursue this issue as such here. It is convenient to consider both exponential and proper dispersion models in the following. In particular, the definitions of residuals illustrate how intuition stemming from the exponential dispersion model case may be useful for arbitrary dispersion models.

Pearson and Wald residuals

Consider a dispersion model $\text{DM}(\mu, \sigma^2)$ with regular unit deviance d and unit variance function V. Let Ω denote the domain for μ, and C the convex support. We define the **Pearson residual** as follows:

$$r_P = \frac{y - \mu}{V^{1/2}(\mu)}.$$

Note that r_P/σ corresponds to standardizing $y - \mu$ using the asymptotic standard deviation $\sigma V^{1/2}(\mu)$ (for σ^2 small). For exponential dispersion models, r_P has mean zero and variance σ^2; this holds only asymptotically in the general case.

Let us define an analogue of the inverse mean value mapping τ^{-1} as follows, for $\mu \in \Omega$,

$$\tau^{-1}(\mu) = \int_{\mu_0}^{\mu} V^{-1}(t)\, dt,$$

where $\mu_0 \in \Omega$ is fixed but arbitrary. We may think of $\theta = \tau^{-1}(\mu)$ as a generalization of the canonical parameter. Based on this, we

define the **Wald residual** as follows:
$$r_W = \left\{\tau^{-1}(y) - \tau^{-1}(\mu)\right\} V^{\frac{1}{2}}(y), \qquad (3.50)$$
for $y \in \Omega$. This definition does not depend on the choice of μ_0. Note that in the discrete case, r_W is generally infinite for y at the boundary of Ω. For a non-steep exponential dispersion model, we may extend the definition of r_W to $y \in C \setminus \Omega$ by replacing y by y_0 in (3.50), where y_0 is the boundary point of Ω nearest to y.

Like the Pearson residual, the Wald residual has asymptotic variance σ^2. We may think of the Wald residual as the standardized difference between y and μ in the scale of the canonical parameter θ. For natural exponential families, r_P^2 and r_W^2 are components of the Pearson and Wald statistics, respectively.

In the following, we refer to r_P and r_W as **crude residuals**.

Score residuals

We now define two types of score residuals that echo the definitions of crude residuals above. The **score residual** is defined by
$$s = -\frac{\partial d}{2\partial \mu} V^{\frac{1}{2}}(\mu).$$
We may think of s as a standardized version of the score statistic. The **dual score residual** is defined by
$$u = \frac{\partial d}{2\partial y} V^{\frac{1}{2}}(y),$$
see Reid (1995). Note that, for exponential dispersion models, s and u reduce to r_P and r_W, respectively. Based on the discussion of the crude residuals above, we may hence think of s and u as measuring differences between y and μ in the scales of the position parameter μ and the canonical parameter θ, respectively.

For the normal distribution, we note that all four residuals above reduce to simply $y - \mu$. In particular, we may think of s and u as two separate ways of obtaining $y - \mu$ by differentiating the unit deviance $(y - \mu)^2$. In the general case, we may justify the score residuals by the formula
$$d(\mu_0 + x\delta; \mu_0 + m\delta) = \frac{\delta^2}{V(\mu_0)} (x - m)^2 + o\left(\delta^2\right), \qquad (3.51)$$
which was derived in Chapter 1. In general s and u are different from each other, but both correspond approximately to a standardized version of the difference $y - \mu$ when y is near μ.

A crucial difference between the score residuals and the crude residuals is their behaviour under transformations. Consider the variate transformation $z = f(y)$ and the corresponding parameter transformation $\xi = f(\mu)$, where f is monotonically increasing and differentiable. As we know from Chapter 1, we need to transform both y and μ by the same transformation in order to preserve the dispersion model structure. However, both s and u are invariant under such transformations, in the sense that if we apply the definitions to the reparametrized deviance

$$d_0(z;\xi) = d\left\{f^{-1}(z); f^{-1}(\xi)\right\}$$

and use the corresponding transformed variance function

$$V_0(\xi) = V\left\{f^{-1}(\xi)\right\}\left[f'\left\{f^{-1}(\xi)\right\}\right]^2,$$

the two score residuals remain unchanged. We call this property **transformation invariance.** This is particularly important outside the exponential dispersion model case, where there is no 'canonical' way of choosing the scale for the observation y. By contrast, the crude residuals are not transformation invariant.

Deviance residuals

A third way to recover $y - \mu$ from the unit deviance is to take its signed square-root, as is obvious from (3.51). We hence define the **deviance residual**, also called the **directed deviance**, as follows:

$$r = \pm d^{\frac{1}{2}}(y;\mu),$$

where $\pm$ denotes the sign of $y - \mu$, so that r_P and r always have the same sign. The deviance residual is clearly transformation invariant. In the following, we refer to the crude, score and deviance residuals as the **standard residuals.**

It follows from (3.51) that all five residuals defined until now are equivalent to each other for y near μ. However, when y is far removed from μ, the five residuals are generally different, and may even have different signs. And while the Pearson residual is a one-to-one function of y for fixed μ, this is not in general the case for the other residuals.

Pierce and Schafer (1986) and McCullagh and Nelder (1989, pp. 37–40) compared the Pearson and deviance residuals for exponential dispersion models, and concluded that the deviance residual is better approximated by the normal distribution than is the Pearson residual. The deviance residual has a bias stemming from the skewness of the distribution, which may be corrected by replacing r by

$$r + \frac{1}{6}\sigma^2 \gamma_3,$$

where $\sigma\gamma_3 = \sigma\kappa_3/\kappa_2^{3/2}$ denotes the skewness (see Section 2.1.3 and Exercise 3.4).

An alternative way of improving the deviance residual, and one that also works outside the exponential dispersion model case, is via the **modified deviance residual** r^*, defined by

$$r^* = \frac{r}{\sigma} + \frac{\sigma}{r}\log\frac{u}{r}$$

provided that u and r have the same sign (Barndorff-Nielsen, 1986; Reid, 1995). Note that the ratio u/r may be expressed as follows:

$$\frac{u}{r} = \frac{\partial r}{\partial y} V^{\frac{1}{2}}(y).$$

Hence, for u and r to have the same sign, r must be a strictly increasing function of y for μ fixed.

The modified deviance residual is transformation invariant, because it is based on the transformation invariant residuals r and u. Note that, contrary to the standard residuals, the modified deviance residual involves σ in its definition. Furthermore, r^* is not in general equivalent to the standard residuals for y near μ.

3.6.2 Tail area approximations

Approximate normality of residuals

The relation with tail area approximations comes from the normal convergence theorem from Chapter 1, by which r_P, and hence all the five standard residuals, are approximately distributed as $N(0, \sigma^2)$ for σ^2 small. The Pearson residual, being a one-to-one function of y for fixed μ, gives rise to the approximation

$$F(y; \mu, \sigma^2) \approx \Phi(r_P/\sigma)$$

for σ^2 small, where $F(y; \mu, \sigma^2)$ denotes the distribution function of $\mathrm{DM}(\mu, \sigma^2)$, and Φ is the standard normal distribution function. Here, and in the following, it is understood that r_P is considered as a function of y for μ fixed.

By the approximate equivalence of the five standard residuals, we may similarly approximate $F(y; \mu, \sigma^2)$ by $\Phi(r_W/\sigma)$, or by $\Phi(u/\sigma)$, and so on. Generally speaking, the more normal a residual is, the better the corresponding approximation is. Furthermore, in residual analysis it is important to use a residual with an approximately normal distribution, because this allows us to judge residual plots based on the assumption that the null distribution of the residuals is normal. Hence, the issues of tail area approximations and normality of residuals are intimately related.

For the modified deviance residual we obtain, as $\sigma^2 \to 0$,

$$F(y; \mu, \sigma^2) = \Phi(r^*) \left\{ 1 + O(\sigma^3) \right\} \qquad (3.52)$$

(Barndorff-Nielsen, 1990; Jensen, 1992). Hence r^* is approximately a standard normal variable, and $\Phi(r^*)$ is yet another approximation to the distribution function $F(y; \mu, \sigma^2)$. As we shall see, this approximation tends to be excellent, and is often significantly better than the closest contender, namely $\Phi(r/\sigma)$ based on the deviance residual.

As may be the case for some of the standard residuals, r^* is not necessarily a one-to-one function of y for μ fixed, see Barndorff-Nielsen (1990). The problem occurs for extreme values of y only, but since one of the main functions of residuals is to identify extreme observations, this may hamper the use of r^* for residual analysis, and may also make the normal approximation to r^* less reliable in the tails of the distribution. The problem seems to occur in cases where u tends to zero while r remains bounded as y tends to the lower terminal of C, which makes r^* tend to infinity, instead of to minus infinity as it ordinarily should. A similar discussion applies for the upper terminal. A dramatic example is the von Mises distribution considered below. Thus, while the approximation based on (3.52) is often very good, it may go badly wrong in some examples.

Lugannani–Rice formula

By a first-order Taylor-expansion of the logarithm we obtain the following approximation for the modified deviance residual:

$$\begin{aligned} r^* &= \frac{r}{\sigma} - \frac{\sigma}{r} \log\left(1 + \frac{r-u}{u}\right) \\ &\approx \frac{r}{\sigma} - \frac{\sigma}{r} \frac{r-u}{u}. \end{aligned}$$

Inserting this in (3.52), and expanding Φ in a first-order Taylor-series, we obtain an alternative version of (3.52),

$$F(y;\mu,\sigma^2) = \Phi^* \left\{1 + O(\sigma^3)\right\},$$

where

$$\Phi^* = \Phi\left(\frac{r}{\sigma}\right) + \sigma\varphi\left(\frac{r}{\sigma}\right)\left(\frac{1}{r} - \frac{1}{u}\right), \qquad (3.53)$$

and where φ denotes the standard normal density. This formula was developed by Lugannani and Rice (1980) in the special case of exponential families, and later generalized. We refer to (3.53) as the **Lugannani–Rice formula.** Depending on the circumstances, Φ^* may be a better approximation to $F(y;\mu,\sigma^2)$ than $\Phi(r^*)$ or vice versa. Note that this approximation is not explicitly based on a definition of residuals. However, the definition

$$r_{LR} = \Phi^{-1}(\Phi^*)$$

gives a residual in the same right as the previous definitions.

In many cases, the approximations based on Φ^* and $\Phi(r^*)$ are asymptotically equivalent at the extremes of C, and the experience so far seems to show that these are precisely the cases where the approximations work well. Barndorff-Nielsen and Cox (1994, p. 212) considered the following set of conditions under which Φ^* and $\Phi(r^*)$ are asymptotically equivalent. The conditions are that $|r| \to \infty$ at the extremes of C and

$$r^{-1}\log(u/r) \to 0, \qquad (3.54)$$

$$u/r^3 \to 0. \qquad (3.55)$$

In particular, these conditions seem to guarantee that the behaviour of r^* at the extremes of C is satisfactory.

For later reference, consider a large negative value of r/σ, and the corresponding Mills' ratio approximation (Read, 1985),

$$\Phi\left(\frac{r}{\sigma}\right) \approx \varphi\left(\frac{r}{\sigma}\right)\frac{\sigma}{-r}.$$

Introducing this approximation in (3.53) gives

$$\Phi^* \approx \frac{-\sigma}{u}\varphi\left(\frac{r}{\sigma}\right). \qquad (3.56)$$

3.6.3 Examples

We now consider a few illustrative examples, and make a numerical and graphical investigation of some of the above approximations.

The examples are important, not just for illustration, but also because ad hoc modifications of the theory are sometimes required.

Gamma distribution

For the gamma distribution, the standard residuals are defined by

$$s = u = r_P = r_W = \frac{y - \mu}{\mu},$$

$$r = \pm\sqrt{2\left(\log\mu/y + y/\mu - 1\right)}.$$

Note that the residuals are functions of y/μ only, so their distribution depends on the value of σ^2 only. Table 3.4 shows the values of the r^* and Lugannani–Rice approximations calculated at the 95% percentile of the $\chi^2(f)$ distribution for small degrees of freedom. While the Lugannani–Rice approximation is slightly more accurate than the r^* approximation in this case, both approximations give good accuracy all the way down to $f = 1$, and both are accurate to four decimal places from $f = 10$ upwards. In terms of the index $\lambda = f/2$ of the gamma distribution, this indicates that the approximation works well for λ as low as 0.5. More elaborate tabulations by Barndorff-Nielsen and Cox (1994, p. 214) for $\lambda = 1, 5$ confirm this.

Figure 3.7 shows some quantile plots for residuals, giving r/σ, r_P/σ and r^* plotted against the normal score $\Phi^{-1}\left\{F(y;\mu,\sigma^2)\right\}$. The perfect approximation corresponds to a curve that follows the diagonal of the plot. The plots confirm the high accuracy of r^*, and r^* is almost indistinguishable from the diagonal. The bias of r mentioned above is clearly illustrated in the plots. The Pearson residual r_P is muchinferior to both r and r^*, and apparently σ^2

Table 3.4. *Values of the* $\Phi(r^*)$ *and Lugannani–Rice* (Φ^*) *approximations for the 95th percentile of the* $\chi^2(f)$ *distribution*

f	$\Phi(r^*)$	Φ^*
1	0.9483	0.9497
2	0.9492	0.9497
3	0.9494	0.9499
4	0.9497	0.9499
5	0.9498	0.9499
7	0.9499	0.9500
10	0.9500	0.9500

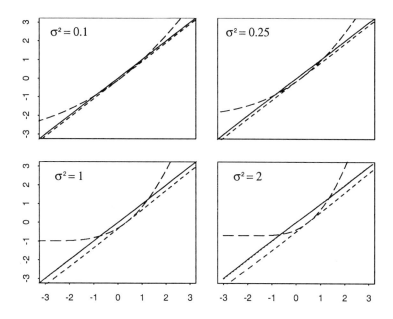

Figure 3.7. *Quantile plots for gamma residuals (r^*: solid line, r/σ: - - -, r_P/σ: — — —, diagonal: $\cdots$). Note that r^* is indistinguishable from the diagonal on these plots.*

must be much smaller than 0.1 before a reasonable approximation in the tails is obtained for this residual.

Simplex distribution

For the simplex distribution $S^-(\mu, \sigma^2)$ from Chapter 1 we obtain the following forms of the score residuals:

$$s = \frac{(y-\mu)\left(\mu^2 + y - 2y\mu\right)}{y(1-y)\mu^{3/2}(1-\mu)^{3/2}}$$

and

$$u = \frac{(y-\mu)\left(y^2 + \mu - 2y\mu\right)}{\sqrt{y(1-y)}\mu^2(1-\mu)^2}.$$

The deviance residual has the form

$$r = \frac{y-\mu}{\sqrt{y(1-y)}\mu(1-\mu)},$$

which goes to infinity at $y = 0$ and 1. We find that

$$\frac{u}{r} = \frac{y^2 + \mu - 2y\mu}{\mu(1-\mu)},$$

and it is easy to check that both (3.54) and (3.55) are satisfied. Hence, we conclude that $\Phi(r^*)$ and Φ^* agree asymptotically for y going to 0 or 1. This suggests that both approximations behave satisfactorily.

von Mises distribution

For the von Mises distribution, the unit deviance has the form

$$d(y;\mu) = 2\left\{1 - \cos(y - \mu)\right\},$$

and the variance function is identically equal to 1. The crude residuals have the form

$$r_P = r_W = y - \mu.$$

It is useful to work with y and μ in $(-\pi, \pi)$ here, and we assume that $y - \mu$ has been reduced modulo 2π to a value in $(-\pi, \pi)$. The score residuals are given by

$$s = u = \sin(y - \mu),$$

and the deviance residual is

$$r = \pm\sqrt{2\left\{1 - \cos(y - \mu)\right\}},$$

where $\pm$ denotes the sign of u.

We now consider the asymptotic behaviour of the residuals as $y - \mu$ tends to $-\pi$. Similar results hold at the upper terminal π. At the lower terminal $-\pi$, we have $u \to 0$, so both $\frac{\sigma}{r}\log\frac{u}{r}$ and r^* tend to infinity. In particular, r^* is not monotone as a function of y.

In spite of this, $\Phi(r^*)$ approximates the von Mises percentiles well in the central range of the distribution. When σ^2 is small, the approximation works well even for very small tail probabilities, as shown by Barndorff-Nielsen and Cox (1994, p. 215). Figure 3.8 confirms that r^* works well in the central range of the distribution, but the non-monotonic behaviour of r^* in the tails is clearly a problem for values of σ^2 around 1 and bigger.

Let us investigate more closely the left-hand tail-behaviour of the distribution. Assume that $\mu = 0$, let $f(y; 0, \sigma^2)$ for $y \in (-\pi, \pi)$ denote the von Mises density function and let $F(y; 0, \sigma^2)$ be the

RESIDUALS AND TAIL AREA APPROXIMATIONS 117

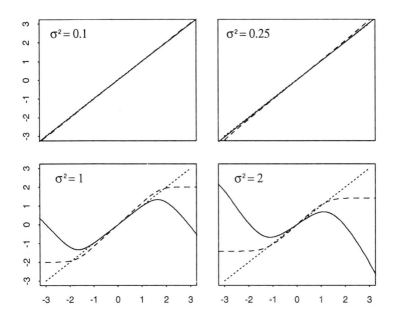

Figure 3.8. *Quantile plots for von Mises residuals (r^*: solid line, r/σ: - - -, diagonal: ···).*

corresponding distribution function. Since f is approximately constant for y near $-\pi$, $F(y; 0, \sigma^2)$ is approximately linear in y, giving

$$F(y; 0, \sigma^2) \approx c(y + \pi), \qquad (3.57)$$

where $c = f(-\pi; 0, \sigma^2) = e^{-\lambda}/\{2\pi I_0(\lambda)\}$ may be approximated by the saddlepoint approximation,

$$c \approx \frac{1}{\sigma}\varphi\left(\frac{2}{\sigma}\right).$$

Note that the approximation (3.56), applying (in principle) for r/σ large negative, gives

$$\Phi^* \approx \frac{-\sigma}{u}\varphi\left(\frac{r}{\sigma}\right) \approx \frac{\sigma}{y+\pi}\varphi\left(\frac{2}{\sigma}\right),$$

in sharp contrast to (3.57).

We now introduce an ad hoc modification of r^* with improved asymptotic behaviour. The modified residual r^+ is designed to give

$$r^+ \approx \Phi^{-1}\{c(y+\pi)\} \approx -\sqrt{-2\log\{c(y+\pi)\}}$$

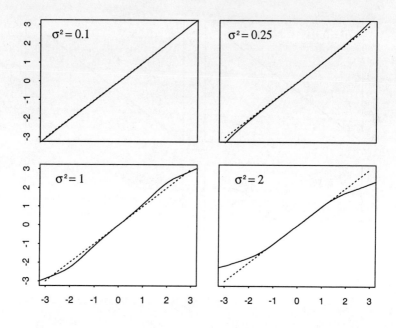

Figure 3.9. *Quantile plots for the ad hoc von Mises residual (r^+: solid line, diagonal: $\cdots$).*

for y near $-\pi$. For y near zero, the modified residual should be approximately r/σ. The interpolation between these two cases is achieved by defining

$$r^+ = \frac{r_0}{\sigma} + \frac{\sigma}{r_0} \log \frac{u}{r_0},$$

where

$$r_0 = \pm \sqrt{\frac{1}{1+\sigma^2} 2 \left\{ 1 - \cos(y - \mu) - \frac{\sigma^2}{1+\sigma^2} \log h(y - \mu) \right\}},$$

and

$$h(y) = \left\{ 2c + \left(\frac{1}{2} - c\right)(1 + \cos y) \right\} \cos \frac{y}{2}.$$

Figure 3.9 shows that the distribution of the residual r^+ is much closer to normal for large values of σ^2 than that of r^*.

Poisson distribution

The distribution function approximations above were all developed for the continuous case, and in the discrete case, approximations to the tail area $1 - F(y; \mu, \sigma^2)$ take on a slightly different form, see e.g. Daniels (1987). However, the following results for the Poisson distribution indicate that the results from the continuous case may be used for integer-valued distributions, not for approximating the tail area as such, but rather as approximations to the symmetrized distribution function

$$G(y) = P(Y < y) + \frac{1}{2} P(Y = y).$$

For the Poisson distribution, the crude and score residuals are identical,

$$r_P = s = \frac{y - \mu}{\sqrt{\mu}}$$

and

$$r_W = u = \sqrt{y} \log \frac{y}{\mu}.$$

The deviance residual is

$$r = \pm \sqrt{2 \left(y \log \frac{y}{\mu} - y + \mu \right)}.$$

Whereas r is finite for $y = 0$, the modified deviance residual is infinite for $y = 0$, where we have $u = 0$ and $r = -\sqrt{2\mu}$. We hence redefine r^* to $r^* = \Phi^{-1}\left\{\frac{1}{2} P(Y = 0)\right\}$ for $y = 0$, so that the corresponding point is located exactly on the identity line.

With this convention for r^* in the case $y = 0$, we illustrate the approximation of $G(y)$ by $\Phi(r^*)$ by plotting r^* as a function of $\Phi^{-1}\{G(y)\}$. Figure 3.10 shows such quantile plots for r and r^*. The plots show that the normal approximation for r^* is very good for expected values as low as $\mu = 0.5$, and slightly better than the normal approximation for r.

3.7 Notes

The idea of an exponential dispersion model may be traced back to Tweedie (1947), who noticed many of the important properties and special cases, and went on to develop another important exponential dispersion model, the inverse Gaussian distribution, which we consider in Chapter 4. However, in contrast to the success of

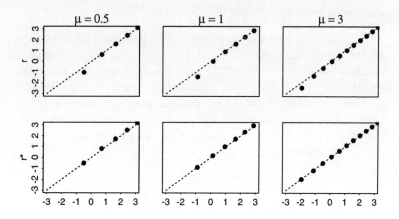

Figure 3.10. *Quantile plots for Poisson residuals r (top) and r* (bottom) (r or r*: •, diagonal: ···).*

Nelder and Wedderburn's ideas, Tweedie's papers remained unnoticed for decades. In particular, Nelder and Wedderburn (1972), whose error distributions for generalized linear models are what we now call reproductive exponential dispersion models, were unaware of Tweedie's paper. Part of the reason for the more ready acceptance of Nelder and Wedderburn's ideas is, of course, that they introduced a class of statistical models, rather than just a class of distributions, and did it at a time when computers were available to carry out the required computations for the statistical analysis. The more systematic study of the mathematical properties of exponential dispersion models was initiated in the papers Jørgensen (1986, 1987a).

A third independent line of development of exponential dispersion models has taken place in the study of exponential families of stochastic processes, see Küchler (1982) and references therein.

The paper by Morris (1982) was seminal for many of the ideas considered here, and his study of quadratic variance functions has stimulated other similar investigations based on the uniqueness theorem for variance functions. In particular, Letac and Mora (1990) showed that there are exactly six strictly cubic variance functions, besides the six quadratic ones found by Morris. Many other characterization results for variance functions and natural exponential families have been developed recently; see e.g. Letac (1992), Kokonendji (1994) and Kokonendji and Seshadri (1994).

EXERCISES   121

The saddlepoint approximation for the distribution of sample averages from the exponential family was introduced by Daniels (1954), and has enjoyed renewed interest following the paper by Barndorff-Nielsen and Cox (1979). The interest in tail-area approximations was much stimulated by the paper by Lugannani and Rice (1980). A survey of these topics was given by Daniels (1987). See also Reid (1988, 1995), Barndorff-Nielsen and Cox (1994, Chapter 6) and references therein.

3.8 Exercises

Exercise 3.1 (Affine transformation) Let Y follow a reproductive exponential dispersion model $ED(\mu, \sigma^2)$.

1. Show that $U = a + bY$, $b \neq 0$, follows a reproductive exponential dispersion model, and find the unit cumulant function and unit variance function of this model.

2. Let Y follow an additive model $ED^*(\theta, \lambda)$. Show that $U = a\lambda + bY$, $b \neq 0$, follows an additive model, and find the unit cumulant function and unit variance function of this model.

Exercise 3.2 (Hyperbolic distribution) The following reproductive model is a special case of the generalized hyperbolic distribution of Barndorff-Nielsen (1978b):

$$p(y; \theta, \lambda) = c(y; \lambda) \exp\left[\lambda \left\{y\theta + (1-\theta^2)^{1/2}\right\}\right], \quad y \in \mathbf{R},$$

where

$$c(y; \lambda) = \frac{\lambda K_1\left(\lambda\sqrt{1+y^2}\right)}{\pi\sqrt{1+y^2}},$$

K_1 being the modified Bessel function with index 1. It is known that this distribution is infinitely divisible.

1. Find Λ, Θ, Ω and C for this distribution. Show that the family is steep.

2. Show that the variance function is

$$V(\mu) = \left(1+\mu^2\right)^{\frac{3}{2}}.$$

3. Find the unit deviance and the saddlepoint approximation.

4. Find the five standard residuals and the modified deviance residual. Derive the $\Phi(r^*)$ and Φ^* approximations. Are the two latter approximations equivalent as $|y| \to \pm\infty$?

5. Show that the limiting form of the distribution for $\lambda \to 0$ is the Cauchy distribution. Hint: The Bessel function satisfies the asymptotic relation $K_1(x) \sim x^{-1}$ as $x \to 0$.

Exercise 3.3 Let Z_1 and Z_2 be independent random variables with distributions

$$Z_i \sim \mathrm{ED}_i^*(\theta, \lambda), \; i = 1, 2,$$

where $\mathrm{ED}_1^*(\theta, \lambda)$ and $\mathrm{ED}_2^*(\theta, \lambda)$ are two continuous additive families, both having the same canonical parameter domain Θ and the same index set Λ. Let the probability density functions of Z_1 and Z_2 be given by

$$p_i(y; \theta, \lambda) = c_i^*(y; \lambda) \exp\{\theta y - \lambda \kappa_i(\theta)\}, \quad y \in \mathbf{R},$$

for $i = 1, 2$.

1. Show that the random variable $Z_+ = Z_1 + Z_2$ follows a continuous additive model $\mathrm{ED}^*(\theta, \lambda)$. Find an expression for the probability density function of Z_+, as explicitly as possible.
2. Find the mean value mapping and mean domain.
3. Find the mean and variance of Z_+.
4. Assume that the two models are identical. Find the unit variance function of $\mathrm{ED}^*(\theta, \lambda)$ in this case.

Exercise 3.4 Show that the ith cumulant for the additive and reproductive cases are

$$\lambda \kappa^{(i)}(\theta) \quad \text{and} \quad \lambda^{1-i} \kappa^{(i)}(\theta),$$

respectively. Find the skewness and kurtosis in the additive and reproductive cases.

Exercise 3.5 Find the probability generating function for a discrete non-negative integer-valued additive model.

Exercise 3.6 (Hermite distribution) The Hermite distribution is a discrete distribution defined by the probability generating function

$$q(u) = \exp\{a_1(u-1) + a_2(u^2-1)\},$$

where $a_1, a_2 > 0$.

1. Show that this distribution is a discrete additive model, and show that it is infinitely divisible. Hint: Find the additive model generated by the distribution given by $a_1 = a_2 = 1$.

2. Show that the unit variance function is, for $\mu > 0$,
$$V(\mu) = \frac{1 + 8\mu - \sqrt{1 + 8\mu}}{4}.$$

3. Show that the distribution converges to the Poisson distribution for $a_2 \to 0$.

4. Show the Hermite distribution may be represented as $Z_1 + 2Z_2$, where Z_1 and Z_2 are independent Poisson variables.

[Kemp and Kemp, 1965]

Exercise 3.7 Show that the standard convolution formula for the χ^2 distribution is a special case of the convolution formula for additive models. Hint: Use the relation
$$\chi^2(f) = Ga^*(-1/2, f/2).$$

Exercise 3.8 Assume that $Z = Z_1 + \cdots + Z_n$, where $Z_1, \ldots, Z_n$ are independent and identically distributed.

1. Show that if $P(|Z| \le c) = 1$ for some constant c, then $\mathrm{var}(Z) \le c^2/n$.

2. Show that an additive model with bounded support cannot be infinitely divisible.

Exercise 3.9 Let $\mathrm{NE}(\mu)$ denote the natural exponential family generated from the uniform distribution $U(0,1)$.

1. Write a computer program that simulates an $\mathrm{NE}(\mu)$ random variable. Hint: Use the rejection method.

2. Use this program to simulate the discrete time additive process corresponding to the additive model generated from $U(0,1)$.

Exercise 3.10 Let $Y \sim \mathrm{Ga}(\mu, \sigma^2)$. Show, by direct calculation, that the density function of the variable $W = (Y - \mu)/\sigma$ converges to the normal density $N(0, \mu^2)$ as $\sigma^2 \to 0$. Hint: Use Stirling's formula $\Gamma(\lambda) \sim (2\pi)^{1/2} \lambda^{\lambda - 1/2} e^{-\lambda}$.

Exercise 3.11 Let $Y \sim \mathrm{ED}(\mu, \sigma^2)$ be a reproductive exponential dispersion model with unit variance function $V(\mu)$, $\mu \in \Omega$.

1. Show that
$$P(|Y - \mu| \ge \sigma) \le V(\mu)$$
for $(\mu, \sigma^{-2}) \in \Omega \times \Lambda$, where Λ is the index set for $\mathrm{ED}(\mu, \sigma^2)$.

2. Show that if $Y \sim \mathrm{Ga}(\mu, \sigma^2)$, then, for $\mu, \sigma^2, \varepsilon > 0$,
$$P\left(\left|\frac{Y}{\mu} - 1\right| \ge \varepsilon\right) \le \frac{\sigma^2}{\varepsilon^2}.$$

Exercise 3.12 Show that $\lambda + \mathrm{Nb}(p, \lambda)$ is an additive model.

Exercise 3.13 (Stirling distribution) Consider the random variable S_n defined by

$$S_n = X_1 + \cdots + X_n - n,$$

where $X_1, X_2, \ldots$ are independent and identically distributed according to the logarithmic distribution in Exercise 2.20 (Chapter 2). The distribution of $X_1 + \cdots + X_n$ is known as the *Stirling distribution*.

1. Find the support of S_n.
2. Show that S_n follows an additive model with index parameter n.
3. Give conditions under which S_n converges to the Poisson distribution as $n \to \infty$.
4. Give conditions under which a standardized version of S_n converges to a normal distribution as $n \to \infty$.
5. Show that the unit variance function of S_n displays the asymptotic behaviour assumed in Theorem 3.5 (at infinity), and find the corresponding gamma convergence result.

[Berg, 1988]

Exercise 3.14 We know that if $Y \sim \mathrm{ED}(\mu, \sigma^2)$, then

$$\frac{Y - \mu}{\sigma} \xrightarrow{d} N\{0, V(\mu)\} \quad \text{as} \quad \sigma^2 \to 0.$$

1. Prove this result using the Central Limit Theorem in connection with the convolution formula.
2. Prove this result by a Taylor-expansion of the cumulant generating function.

Exercise 3.15 Let X and Y be independent and identically distributed according to the Poisson distribution $\mathrm{Po}(\lambda)$, where $\lambda > 0$. Let $c(z; \lambda)$ denote the probability function for $Z = X - Y$.

1. Find an expression for c, and find its cumulant generating function $K(s; \lambda)$.
2. Consider the natural exponential family generated by Z, with probability function

$$p(z; \theta, \lambda) = c(z; \lambda) \exp\{\theta z - K(\theta; \lambda)\}, \; z \in \mathbf{Z}.$$

Find the domain for (θ, λ), and show that $p(z; \theta, \lambda)$ is an additive family $\mathrm{ED}^*(\theta, \lambda)$.

3. Find the expectation and the mean domain and show that the model is steep.

4. Show that the unit variance function is $V(\mu) = \sqrt{\mu^2 + 4}$. Hint: The following formula may be useful:
$$\frac{1}{\sqrt{a}+b} = \frac{\sqrt{a}-b}{a-b^2}.$$

Exercise 3.16 Let $\mathrm{ED}_i^*(\theta_i, \lambda_i)$, $i = 1, 2$, denote two additive models, with probability density functions
$$p_i(z; \theta_i, \lambda_i) = c_i(z; \lambda_i) \exp\{\theta_i z - \lambda_i \kappa_i(\theta_i)\},$$
$i = 1, 2$. Assume that $\kappa_1(\Theta_1) \subseteq \Theta_2$, where Θ_1 and Θ_2 are the respective canonical parameter domains for the two models, and assume that $\mathrm{ED}_2^*(\theta_2, \lambda_2)$ is a non-negative distribution. Let Z_1 and Z_2 be random variables such that $Z_2 \sim \mathrm{ED}_2^*\{\kappa_1(\theta_1), \lambda\}$, and let the conditional distribution of Z_1 given $Z_2 = z_2$ be $\mathrm{ED}_1^*(\theta_1, z_2)$, defined by the conditional probability density function
$$p(z_1|z_2; \theta_1) = c_1(z_1; z_2) \exp\{\theta_1 z_1 - z_2 \kappa_1(\theta_1)\}.$$
Here we use the convention that $\mathrm{ED}_1^*(\theta_1, 0)$ denotes the degenerate distribution in 0, with $c_1(z_1; 0)$ defined by
$$c_1(z_1; 0) = \begin{cases} 1 & \text{for} \quad z_1 = 0 \\ 0 & \text{for} \quad z_1 > 0. \end{cases}$$

1. Show that the marginal distribution of Z_1 is an additive model $\mathrm{ED}^*(\theta_1, \lambda)$ with unit cumulant function $\kappa_2\{\kappa_1(\theta_1)\}$, $\theta_1 \in \Theta_1$.

2. Find the expectation and variance of Z_1.

3. Suppose that $\mathrm{ED}_1^*(\theta_1, \lambda_1)$ and $\mathrm{ED}_2^*(\theta_2, \lambda_2)$ are the binomial and the Poisson distributions, respectively. Find the marginal distribution of Z_1 in this case.

Exercise 3.17 Find the exponential dispersion model corresponding to the unit variance function $V(\mu) = \mu^2$, $\mu < 0$.

Exercise 3.18 Find the exponential dispersion model corresponding to the unit variance function $V(\mu) = a + b\mu$.

Exercise 3.19 Let the probability function $c(z)$ be defined by
$$c(z) = \begin{cases} 1/4 & \text{for} \quad z = \pm 1 \\ 1/2 & \text{for} \quad z = 0. \end{cases}$$

1. Find the natural exponential family generated by $c(z)$.

2. Show that its variance function is $V(\mu) = (1 - \mu^2)/2$, $|\mu| < 1$.

3. Determine which of Morris' six categories this function belongs to, and explain the relation with the corresponding elementary distribution.

Exercise 3.20 Define $V(\mu) = (\mu - 1)/2$, for $\mu > 1$. Show that the function
$$\xi \mapsto \lambda V(\xi/\lambda) = \frac{\xi - \lambda}{2}$$
is the variance function of a discrete exponential dispersion model $ED^*(\theta, \lambda)$, and find the corresponding probability function.

Exercise 3.21 Carry out in detail the transformation
$$z = \frac{1}{\pi} \log \frac{u}{1-u}$$
leading from (3.38) to (3.37).

Exercise 3.22 Prove the result
$$c^{-1}\text{Nb}\left(\frac{c\mu}{1+c\mu}, \lambda\right) \xrightarrow{d} \text{Ga}(\mu, \lambda^{-1}) \quad \text{as} \quad c \to \infty.$$
using Theorem 3.5. Hint: Apply the duality transformation to the negative binomial distribution to obtain a reproductive model to which the theorem may be applied.

Exercise 3.23 Find the skewness and kurtosis for the generalized hyperbolic secant distribution.

Exercise 3.24 Find the saddlepoint approximation to the generalized hyperbolic secant distribution.

Exercise 3.25 Investigate the Lugannani–Rice and r^* approximations for the binomial and negative binomial distributions, along the same lines as done for the Poisson distribution in Section 3.6.3.

CHAPTER 4

Tweedie models

We now study the class of Tweedie exponential dispersion models, which are exponential dispersion models closed under scale transformation or under translations.

4.1 Characterization and properties

We begin by characterizing the Tweedie models in terms of scale transformations, and consider their main properties. Models closed under translations will be considered in Section 4.5.

4.1.1 Scale transformations

The main class of Tweedie unit variance functions is defined as follows:
$$V_p(\mu) = \mu^p, \quad \mu \in \Omega_p, \tag{4.1}$$
where $p \in \mathbf{R}$. The reproductive exponential dispersion model corresponding to (4.1), if it exists, is denoted $Y \sim \text{Tw}_p(\mu, \sigma^2)$. By definition, this model has mean μ and variance
$$\text{var}\, Y = \sigma^2 \mu^p, \quad \mu \in \Omega_p. \tag{4.2}$$
We have already met some examples of Tweedie models, namely the normal ($p = 0$), Poisson ($p = 1$) and gamma ($p = 2$) distributions. The relation between the new notation $\text{Tw}_p(\mu, \sigma^2)$ and the notation for these distributions is as follows:
$$\text{Tw}_0(\mu, \sigma^2) = N(\mu, \sigma^2),$$
$$\text{Tw}_1(\mu, \sigma^2) = \sigma^2 \text{Po}\left(\mu/\sigma^2\right),$$
$$\text{Tw}_2(\mu, \sigma^2) = \text{Ga}(\mu, \sigma^2).$$
The formula for the Poisson distribution follows from the duality transformation, by which the reproductive form of the Poisson distribution is $\sigma^2 \text{Po}(\mu/\sigma^2)$.

Both the normal and the gamma families are closed with respect to scale transformations,
$$cN(\mu,\sigma^2) = N(c\mu, c^2\sigma^2),$$
$$c\operatorname{Ga}(\mu,\sigma^2) = \operatorname{Ga}(c\mu,\sigma^2).$$
So is the Poisson family in its reproductive form (see Exercise 4.2),
$$c\operatorname{Tw}_1(\mu,\sigma^2) = \operatorname{Tw}_1(c\mu, c\sigma^2).$$
The next theorem shows that all exponential dispersion models that are closed with respect to scale transformations are Tweedie models.

Theorem 4.1 *Let* $\operatorname{ED}(\mu,\sigma^2)$ *denote a reproductive exponential dispersion model such that* $1 \in \Omega$ *and* $V(1) = 1$. *If the model is closed with respect to scale transformations, such that there exists a function* $f : \mathbf{R}_+ \times \Lambda^{-1} \to \Lambda^{-1}$ *for which*
$$c\operatorname{ED}(\mu,\sigma^2) = \operatorname{ED}\left\{c\mu, f(c,\sigma^2)\right\} \quad \forall c > 0, \tag{4.3}$$
then:

1. $\operatorname{ED}(\mu,\sigma^2)$ *is a Tweedie model for some* $p \in \mathbf{R}$;
2. $f(c,\sigma^2) = c^{2-p}\sigma^2$;
3. *The mean domain is* $\Omega = \mathbf{R}$ *for* $p = 0$ *and* $\Omega = \mathbf{R}_+$ *for* $p \neq 0$;
4. *The model is infinitely divisible.*

Proof. From $Y \sim \operatorname{ED}(\mu,\sigma^2)$ we obtain
$$\operatorname{var} cY = c^2\sigma^2 V(\mu), \tag{4.4}$$
whereas (4.3) gives
$$\operatorname{var} cY = f(c,\sigma^2) V(c\mu). \tag{4.5}$$
Taking $\sigma^2 = 1$ and letting $g(c) = c^2/f(c,1)$, we obtain the relation
$$V(c\mu) = g(c)V(\mu) \quad \forall c > 0,\ \mu \in \Omega. \tag{4.6}$$
Since this relation holds for all $c > 0$, we have either $\Omega = \mathbf{R}$ or, since $1 \in \Omega$, $\Omega = \mathbf{R}_+$. Straightforward arguments (see Exercise 4.1) show that the function V satisfying (4.6) is
$$V(\mu) = \mu^p,\ \mu \in \Omega,$$
for some $p \in \mathbf{R}$. This also shows that $\Omega = \mathbf{R}$ for $p = 0$ and $\Omega = \mathbf{R}_+$ for $p \neq 0$. By comparing (4.4) and (4.5) for general σ^2, we find $f(c,\sigma^2) = c^{2-p}\sigma^2$. For $p \neq 2$, this implies that $\Lambda = \mathbf{R}_+$, because c varies in $\mathbf{R}_+$, and hence the dispersion parameter $c^{2-p}\sigma^2$

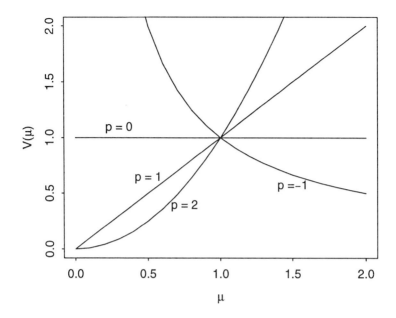

Figure 4.1. *Some Tweedie unit variance functions.*

varies in $\mathbf{R}_+$, showing infinite divisibility. For $p = 2$, the same conclusion follows because then $V(\mu) = \mu^2$, corresponding to the gamma distribution, which is infinitely divisible. □

Theorem 4.1 does not provide information about which values of p are possible. We show in Proposition 4.2 below that only the following values of p correspond to exponential dispersion models:

$$p \in (-\infty, 0] \cup [1, \infty).$$

Figure 4.1 shows some examples of Tweedie variance functions. Table 4.1 summarizes the class of Tweedie models. Most of the models are continuous, except for the Poisson ($p = 1$) and compound Poisson models ($1 < p < 2$), the latter being continuous with a positive probability in zero. The case $p = \infty$ in the table refers to the models with exponential unit variance functions to be considered in Section 4.5.

By the result $f(c, \sigma^2) = c^{2-p}\sigma^2$ from Theorem 4.1, we find that the scale transformation property (4.3) has the following form:

$$c \text{Tw}_p(\mu, \sigma^2) = \text{Tw}_p(c\mu, c^{2-p}\sigma^2). \tag{4.7}$$

Table 4.1. *Summary of Tweedie exponential dispersion models*

Distributions	p	S	Ω	Θ
Extreme stable	$p < 0$	$\mathbf{R}$	$\mathbf{R}_+$	$\mathbf{R}_0$
Normal	$p = 0$	$\mathbf{R}$	$\mathbf{R}$	$\mathbf{R}$
[Do not exist]	$0 < p < 1$	—	$\mathbf{R}_+$	$\mathbf{R}_0$
Poisson	$p = 1$	$\mathbf{N}_0$	$\mathbf{R}_+$	$\mathbf{R}$
Compound Poisson	$1 < p < 2$	$\mathbf{R}_0$	$\mathbf{R}_+$	$\mathbf{R}_-$
Gamma	$p = 2$	$\mathbf{R}_+$	$\mathbf{R}_+$	$\mathbf{R}_-$
Positive stable	$2 < p < 3$	$\mathbf{R}_+$	$\mathbf{R}_+$	$-\mathbf{R}_0$
Inverse Gaussian	$p = 3$	$\mathbf{R}_+$	$\mathbf{R}_+$	$-\mathbf{R}_0$
Positive stable	$p > 3$	$\mathbf{R}_+$	$\mathbf{R}_+$	$-\mathbf{R}_0$
Extreme stable	$p = \infty$	$\mathbf{R}$	$\mathbf{R}$	$\mathbf{R}_-$

Notation: $-\mathbf{R}_0 = (-\infty, 0]$

This formula generalizes the scale transformation properties of the normal, gamma and Poisson distributions mentioned above. By the duality transformation, the additive form of the model is hence

$$\operatorname{Tw}_p^*(\theta, \lambda) = \operatorname{Tw}_p\left\{\lambda \tau_p(\theta), \lambda^{1-p}\right\}, \tag{4.8}$$

where τ_p denotes the mean value mapping. The Tweedie models hence share with the normal and gamma models the property that they have both an additive and a reproductive form. The inverse of the relation (4.8), expressing the additive form in terms of the reproductive one, is as follows:

$$\operatorname{Tw}_p(\mu, \sigma^2) = \operatorname{Tw}_p^*\left\{\tau_p^{-1}\left(\mu \sigma^{\frac{2}{p-1}}\right), \sigma^{\frac{2}{p-1}}\right\},$$

see Exercise 4.5.

4.1.2 Cumulant generating function

The first step in our investigation of the Tweedie class is to derive its cumulant generating function. As we know from the characterization theorem for variance functions, an exponential dispersion model is uniquely characterized by its unit variance function, and we now follow the steps of the proof of the characterization theorem in order to find the actual form of the moment generating function. The formulas in the following are derived for a general value of $p \in \mathbf{R}$, since we have not yet proved anything about the domain for p.

CHARACTERIZATION AND PROPERTIES

It is convenient to introduce the parameter α, defined by

$$\alpha = \frac{p-2}{p-1}, \tag{4.9}$$

with inverse relation

$$p = \frac{\alpha - 2}{\alpha - 1}. \tag{4.10}$$

A further useful relation is

$$(p-1)(\alpha - 1) = -1.$$

We let $\alpha = -\infty$ correspond to the limiting case $p = 1$. In this way (4.9) and (4.10) define a one-to-one relation between $p \in \mathbf{R}$ and $\alpha \in [-\infty, 1) \cup (1, \infty)$. We use α and p interchangeably in the following.

We let $V_p(\mu) = \mu^p$, for $\mu \in \Omega_p$, denote the Tweedie unit variance function, and we let κ_p and τ_p, with domain Θ_p denote the corresponding unit cumulant function and mean value mapping, respectively.

To find the exponential dispersion model corresponding to V_p, if it exists, we must solve the following two differential equations:

$$\frac{\partial \tau_p^{-1}}{\partial \mu} = \mu^{-p}, \quad \mu \in \Omega_p \tag{4.11}$$

and, in turn,

$$\kappa_p'(\theta) = \tau_p(\theta), \quad \theta \in \Theta_p. \tag{4.12}$$

From (4.11) we find, for $\theta \in \Theta_p$,

$$\tau_p(\theta) = \begin{cases} \left(\frac{\theta}{\alpha - 1}\right)^{\alpha - 1} & \text{for } p \neq 1 \\ e^\theta & \text{for } p = 1. \end{cases}$$

From τ_p we find κ_p by solving (4.12), which gives, for $\theta \in \Theta_p$,

$$\kappa_p(\theta) = \begin{cases} \frac{\alpha - 1}{\alpha}\left(\frac{\theta}{\alpha - 1}\right)^\alpha & \text{for } p \neq 1, 2 \\ -\log(-\theta) & \text{for } p = 2 \\ e^\theta & \text{for } p = 1. \end{cases} \tag{4.13}$$

In both (4.11) and (4.12), we have ignored the arbitrary constants in the solutions, which do not affect the results.

The canonical parameter domain Θ_p is the largest interval for which κ_p is finite, whence

$$\Theta_p = \begin{cases} \mathbf{R} & \text{for} \quad p = 0, 1 \\ \mathbf{R}_0 & \text{for} \quad p < 0 \text{ or } 0 < p < 1 \\ \mathbf{R}_- & \text{for} \quad 1 < p \le 2 \\ (-\infty, 0] & \text{for} \quad 2 < p < \infty. \end{cases} \quad (4.14)$$

If an exponential dispersion model corresponding to (4.13) exists, the cumulant generating function of the corresponding convolution model is, for $s \in \Theta_p - \theta$,

$$K_p^*(s; \theta, \lambda) = \begin{cases} \lambda \kappa_p(\theta) \left\{ \left(1 + \frac{s}{\theta}\right)^\alpha - 1 \right\} & \text{for} \quad p \ne 1, 2 \\ -\lambda \log\left(1 + \frac{s}{\theta}\right) & \text{for} \quad p = 2 \\ \lambda e^\theta (e^s - 1) & \text{for} \quad p = 1. \end{cases} \quad (4.15)$$

The cumulant generating function for the reproductive form of the distribution, parametrized as $\mathrm{Tw}_p \{\tau_p(\theta), 1/\lambda\}$, is then $K_p(s; \theta, \lambda) = K_p^*(s/\lambda; \theta, \lambda)$, or

$$K_p(s; \theta, \lambda) = \begin{cases} \lambda \kappa_p(\theta) \left\{ \left(1 + \frac{s}{\theta \lambda}\right)^\alpha - 1 \right\} & \text{for} \quad p \ne 1, 2 \\ -\lambda \log\left(1 + \frac{s}{\theta \lambda}\right) & \text{for} \quad p = 2 \\ \lambda e^\theta \left\{ \exp\left(\frac{s}{\lambda}\right) - 1 \right\} & \text{for} \quad p = 1. \end{cases} \quad (4.16)$$

We let $\mathrm{Tw}_p(0, \sigma^2)$ (for $p \le 0$) and $\mathrm{Tw}_p(\infty, \sigma^2)$ (for $p > 2$) denote the limiting cases of (4.16) for $\theta = 0$.

To decide whether or not (4.15) and (4.16) are moment generating functions, we first rule out the case $0 < p < 1$. The remaining cases, corresponding to $p \notin (0, 1)$, all correspond to exponential dispersion models, as we show in Section 4.2.

Proposition 4.2 *There are no exponential dispersion models with unit variance functions $V(\mu) = \mu^p$ for $0 < p < 1$.*

Proof. Let $p \in (0, 1)$ be given and let $\alpha > 2$ be the corresponding value of α. Suppose that $K_p^*(s; \theta, \lambda)$ from equation (4.15) is a cumulant generating function of an additive exponential dispersion model $\mathrm{Tw}_p^*(\theta, \lambda)$. This model has canonical parameter domain $\Theta_p = \mathbf{R}_0$, so $0 \in \overline{\Theta}_p$. The variance of the distribution $\mathrm{Tw}_p^*(\theta, \lambda)$ is

$$\lambda \kappa_p''(\theta) = \lambda \left(\frac{\theta}{\alpha - 1}\right)^{\alpha - 2},$$

which is zero for $\theta = 0$ because $\alpha > 2$. Hence $\mathrm{Tw}_p^*(0, \lambda)$ is a degenerate distribution with moment generating function e^{sc} for some $c \in \mathbf{R}$, which contradicts (4.15). We conclude that $K_p^*(s; \theta, \lambda)$ is not a moment generating function in the case $0 < p < 1$. $\square$

CHARACTERIZATION AND PROPERTIES

Table 4.2. *Unit deviances for Tweedie models*

Model	Unit deviance
$N(\mu, \sigma^2)$ $(p=0)$	$(y-\mu)^2$
$\text{Po}(\mu)$ $(p=1)$	$2\{y\log(y/\mu) - (y-\mu)\}$
$\text{Ga}(\mu, \sigma^2)$ $(p=2)$	$2\{\log(\mu/y) + y/\mu - 1\}$
$\text{Tw}_p(\mu, \sigma^2)$ $(p \neq 0,1,2)$	$2\left\{\frac{[\max\{y,0\}]^{2-p}}{(1-p)(2-p)} - \frac{y\mu^{1-p}}{1-p} + \frac{\mu^{2-p}}{2-p}\right\}$
$\text{Tw}_\infty(\mu, \sigma^2, \beta)$	$2\beta^{-2}\left\{e^{-\beta y} + e^{-\beta\mu}(\beta y - \beta\mu - 1)\right\}$

In the case where the support is contained in $\mathbf{R}_+$, Theorem 2.15 shows that the right derivative $V'(0^+)$ is finite, ruling out the values $1 < p < 2$. It seems plausible that the condition $V(0^+) = 0$ implies that the support is contained in $\mathbf{R}_+$, but this was not proved in Theorem 2.15.

By comparing the third and fourth columns of Table 4.1, we find that the Tweedie models are steep in the cases $p=0$ and $p \geq 1$. The models are not steep in the case $p < 0$, where the support is $\mathbf{R}$ and the mean domain is $\Omega = \mathbf{R}_+$.

4.1.3 Unit deviance

We now consider the unit deviances for the Tweedie models, summarized in Table 4.2. The Poisson, gamma and normal cases have already been considered in Chapter 3. We hence consider the case $p \neq 0, 1, 2$, where the unit cumulant function is

$$\kappa_p(\theta) = \frac{\alpha-1}{\alpha}\left(\frac{\theta}{\alpha-1}\right)^\alpha.$$

To calculate $\sup_\theta \{y\theta - \kappa_p(\theta)\}$, we need to solve the equation

$$y = \left(\frac{\theta}{\alpha-1}\right)^{\alpha-1}. \tag{4.17}$$

This equation has a solution in $\text{int}\,\Theta_p$ only if $y > 0$ while if $y \leq 0$, the derivative of $\theta y - \kappa_p(\theta)$ is negative. Hence the supremum is

$$\frac{\{\max(y,0)\}^{2-p}}{(1-p)(2-p)}.$$

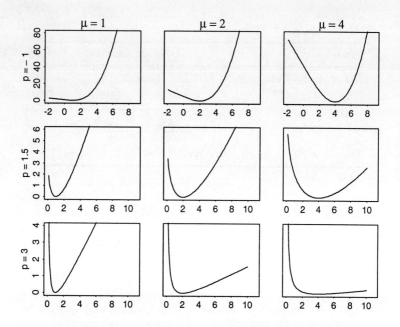

Figure 4.2. *Some Tweedie unit deviances.*

The unit deviance hence takes the following form for $\mu \in \Omega_p$:

$$d_p(y;\mu) = 2\left[\frac{\{\max(y,0)\}^{2-p}}{(1-p)(2-p)} - \frac{y\mu^{1-p}}{1-p} + \frac{\mu^{2-p}}{2-p}\right].$$

Note that in the cases $p < 0$ and $1 < p < 2$, the unit deviance is finite for $y = 0$, with value

$$d_p(0;\mu) = \frac{\mu^{2-p}}{2-p}.$$

Figure 4.2 shows some plots of Tweedie unit deviances. This plot indicates that the value of p is related to the asymmetry of the unit deviance. Note in particular that the unit deviance is very asymmetric in the cases $p = 3$ and $p = -1$.

4.2 Special cases

Before considering the stable Tweedie models we consider the definition of stable distributions.

4.2.1 Stable distributions

Let $X_1, \ldots, X_n$ be independent and identically distributed with non-degenerate distribution F. If, for all n, there exist constants b_n and c_n, such that

$$b_n + c_n(X_1 + \cdots + X_n) \sim F, \qquad (4.18)$$

then F is said to be **stable**. If $b_n = 0$ for all n, then F is said to be **strictly stable**. The constants c_n are given by $c_n = n^{-1/\alpha}$ for some $0 < \alpha \leq 2$ (see Feller, 1971, Chapter 6). The parameter α is called the **index** of F (sometimes called the stability index, in order to avoid confusion with the index parameter of exponential dispersion models). All stable distributions are continuous.

Note that if X is a stable random variable, then (4.18) may be written as

$$X = \sum_{i=1}^{n} \left(\frac{b_n}{n} + c_n X_i \right),$$

which shows that any stable distribution is infinitely divisible. The converse is not true; for example, the gamma distribution is infinitely divisible, but not stable.

The class of stable distributions is parametrized by four parameters, namely location, scale, index, and a fourth parameter β with domain $[-1, 1]$. Strictly stable distributions do not have any location parameter. **Extreme stable distributions** are defined as the special case where the parameter β is either $+1$ or -1. Extreme stable distributions with $\alpha < 1$ have support $\mathbf{R}_+$ or $\mathbf{R}_-$ and are strictly stable. In the positive case they are called **positive stable distributions**. All other types of extreme stable distributions have support $\mathbf{R}$.

Most stable distributions, like the Cauchy (see Exercise 2.3) do not generate natural exponential families, because they have degenerate moment generating functions. In particular, except for the normal case, moments of order bigger than α do not exist for stable variables. However, positive and extreme stable distributions have moment generating functions that are powers of θ, see Eaton, Morris and Rubin (1971). We show next that they appear as special cases of Tweedie models.

4.2.2 Stable Tweedie models

We now consider the cases $p \leq 0$ and $p > 2$ of the Tweedie class, and show that they correspond to extreme stable distributions.

Note first that by (4.14) we have $0 \in \Theta_p$ in these cases, so that we may consider the distribution $\mathrm{Tw}_p^*(0,\lambda)$.

Theorem 4.3 *The distribution $\mathrm{Tw}_p^*(0,\lambda)$ is strictly stable with index $1 < \alpha \leq 2$ if $p \leq 0$ and index $0 < \alpha < 1$ if $p > 2$, where $\alpha = (p-2)/(p-1)$.*

Proof. Taking $\theta = 0$ in (4.15), we find that the cumulant generating function of $\mathrm{Tw}_p^*(0,\lambda)$ is

$$K_p^*(s;0,\lambda) = \lambda\{\kappa_p(s) - \kappa_p(0)\} = \lambda\frac{\alpha-1}{\alpha}\left(\frac{s}{\alpha-1}\right)^\alpha.$$

If $X_1,\ldots,X_n$ are independent with distribution $\mathrm{Tw}_p^*(0,\lambda)$, then $X_1 + \cdots + X_n$ has cumulant generating function

$$\begin{aligned}
nK_p^*(s;0,\lambda) &= n\lambda\frac{\alpha-1}{\alpha}\left(\frac{s}{\alpha-1}\right)^\alpha \\
&= \lambda\frac{\alpha-1}{\alpha}\left(\frac{sn^{1/\alpha}}{\alpha-1}\right)^\alpha \\
&= K_p^*\left(sn^{1/\alpha};0,\lambda\right).
\end{aligned}$$

Hence $X_1 + \cdots + X_n$ and $n^{1/\alpha}X_1$ have the same distribution, which shows that the distribution $\mathrm{Tw}_p^*(0,\lambda)$ is strictly stable with index α. □

The theorem shows that, for each λ fixed, the natural exponential family $\mathrm{Tw}_p^*(\Theta_p,\lambda)$ is generated by the stable distribution $\mathrm{Tw}_p^*(0,\lambda)$. The two models $\mathrm{Tw}_p^*(\theta,\lambda)$ and $\mathrm{Tw}_p(\mu,\sigma^2)$ are hence referred to as **stable Tweedie exponential dispersion models**. In the reproductive notation, the model $\mathrm{Tw}_p(0,\sigma^2)$ is an extreme stable distribution in the case $p \leq 0$, and the model $\mathrm{Tw}_p(\infty,\sigma^2)$ is a positive stable distribution in the case $p > 2$.

It is important to note that the members of the stable Tweedie models are *not* stable as such for $\theta \neq 0$. However, this terminology seems reasonable in view of the fact that these models have many properties similar to stable distributions. In particular, they appear as limiting distributions (as do all the Tweedie models) in a kind of generalized central limit theorem, see Section 4.4.2.

We now consider the probability density functions of the stable Tweedie models, which may be expressed as infinite series. The positive stable distributions $\mathrm{Tw}_p^*(0,\lambda)$ for $p > 2$ have probability

SPECIAL CASES

density functions $c_p^*(z; \lambda)$, given for $z > 0$ by

$$c_p^*(z; \lambda) = \frac{1}{\pi z} \sum_{k=1}^{\infty} \frac{\Gamma(1 + \alpha k)}{k!} \lambda^k \kappa_p^k \left(-\frac{1}{z}\right) \sin(-k\pi\alpha). \quad (4.19)$$

The extreme stable distributions with $p < 0$ have density functions for $z \in \mathbf{R}$ given by

$$c_p^*(z; \lambda) = \frac{1}{\pi z} \sum_{k=1}^{\infty} \frac{\Gamma\left(1 + \frac{k}{\alpha}\right)(-z)^k}{k! \, \lambda^{\frac{k}{\alpha}} \kappa_p^{\frac{k}{\alpha}}(1)} \sin\left(\frac{-k\pi}{\alpha}\right). \quad (4.20)$$

These results may be proved by Fourier inversion, see Feller (1971, p. 581).

By the duality transformation, the reproductive exponential dispersion models $\mathrm{Tw}_p(\mu, \sigma^2)$ have probability density functions

$$p(y; \theta, \lambda) = \lambda c_p^*(\lambda y; \lambda) \exp[\lambda\{\theta y - \kappa_p(\theta)\}]. \quad (4.21)$$

These models account for all values of α between 0 and 2, except for $\alpha = 1$. The latter case will be considered in Section 4.5. Figure 4.3 shows some reproductive Tweedie densities in the case $p = -1$.

4.2.3 Inverse Gaussian distribution

The simplest special case of positive stable Tweedie distributions is the inverse Gaussian distribution, which corresponds to $p = 3$. We begin by considering the inverse Gaussian distribution function.

Let Φ denote the standard normal distribution function, let $\xi \geq 0$, $\lambda > 0$, and define the function $F(\,\cdot\,; \xi, \lambda)$ as

$$F(y; \xi, \lambda) = \Phi\left(\xi\sqrt{\lambda y} - \sqrt{\lambda/y}\right) + e^{2\lambda\xi} \Phi\left(-\xi\sqrt{\lambda y} - \sqrt{\lambda/y}\right)$$

for $y > 0$ and as 0 for $y \leq 0$. Then, as proved in Exercise 4.6, $F(\,\cdot\,; \xi, \lambda)$ is a continuous distribution function, corresponding to the probability density function defined for $y > 0$ by

$$F'(y; \xi, \lambda) = \sqrt{\frac{\lambda}{2\pi y^3}} \exp\left\{-\frac{\lambda}{2}\left(\xi^2 y + y^{-1} - 2\xi\right)\right\}. \quad (4.22)$$

This defines the **inverse Gaussian distribution**.

To show that (4.22) is an exponential dispersion model, we introduce the parameter $\theta = -\xi^2/2 \leq 0$ and rewrite (4.22) in the

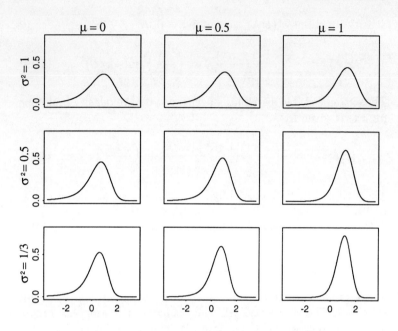

Figure 4.3. *Some Tweedie densities for $p = -1$ ($\alpha = 3/2$).*

form

$$p(y; \theta, \lambda) = \sqrt{\frac{\lambda}{2\pi y^3}} \exp\left[-\frac{\lambda}{2y} + \lambda\left\{\theta y + (-2\theta)^{1/2}\right\}\right]. \quad (4.23)$$

This is a reproductive exponential dispersion model with unit cumulant function $\kappa(\theta) = -(-2\theta)^{1/2}$, mean

$$\mu = (-2\theta)^{-\frac{1}{2}}$$

and mean domain $\Omega = \mathbf{R}_+$, the model being steep. The corresponding unit variance function is easily seen to be of the form claimed, namely

$$V(\mu) = \mu^3, \quad \mu > 0.$$

We denote the distribution by $\mathrm{IG}(\mu, \sigma^2) = \mathrm{Tw}_3(\mu, \sigma^2)$, where $\sigma^2 = 1/\lambda$. Figure 4.4 shows some inverse Gaussian density functions. Note again the considerable skewness of these distributions, which we already noted in connection with the unit deviances in Figure 4.2.

SPECIAL CASES

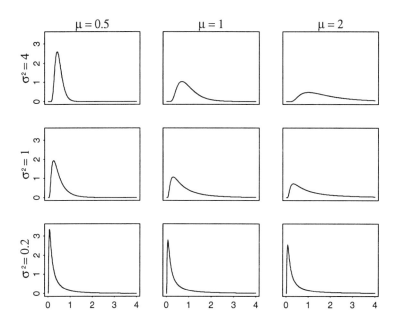

Figure 4.4. *Some inverse Gaussian density finctions.*

Writing the density (4.23) in standard form, for $y > 0$,

$$p(y; \mu, \sigma^2) = \frac{1}{\sqrt{2\pi\sigma^2}} y^{-\frac{3}{2}} \exp\left\{-\frac{1}{2\sigma^2}\frac{(y-\mu)^2}{y\mu^2}\right\}. \qquad (4.24)$$

we find that the saddlepoint approximation is exact, just as for the normal distribution.

In the case $\theta = 0$ ($\mu = \infty$), the density (4.23) becomes

$$p(y; 0, \lambda) = \sqrt{\frac{\lambda}{2\pi y^3}} \exp\left(-\frac{\lambda}{2y}\right). \qquad (4.25)$$

which is the density of the positive stable distribution with index $\alpha = 1/2$, denoted $\mathrm{IG}(\infty, \sigma^2)$. Both mean and variance are infinite in this case.

The scale transformation property of the inverse Gaussian distribution takes the form

$$cY \sim \mathrm{IG}\left(c\mu, \sigma^2/c\right).$$

By taking $c = \lambda$, $\mu = \tau(\theta)$ and $\sigma^2 = 1/\lambda$ in this formula we obtain the duality transformation, giving the following additive form of

the distribution:
$$IG^*(\theta, \lambda) = IG\left(\frac{\lambda}{\sqrt{-2\theta}}, \frac{1}{\lambda^2}\right).$$

4.2.4 Compound Poisson models

We now investigate the case $\alpha < 0$, corresponding to $1 < p < 2$, and show that it corresponds to a class of compound Poisson distributions.

Let $N, X_1, X_2, \ldots$ denote a sequence of independent random variables, such that N is Poisson distributed $\text{Po}(m)$ and the X_is are identically distributed. Define

$$Z = \sum_{i=1}^{N} X_i \quad (4.26)$$

where Z is defined as 0 for $N = 0$. The distribution (4.26) is called a **compound Poisson distribution**. Now assume that

$$m = \lambda \kappa_p(\theta)$$

and
$$X_i \sim \text{Ga}^*(\theta, -\alpha), \quad i = 1, 2, \ldots, \quad (4.27)$$

where $\alpha, \theta < 0$ and where α and p are related by (4.9), as above. Note that, by the convolution formula, we have

$$Z|N = n \sim \text{Ga}^*(\theta, -n\alpha), \quad (4.28)$$

for $n \geq 1$.

Let $M^*(s; \theta, \lambda) = (1 + s/\theta)^{-\lambda}$ denote the moment generating function of the distribution $\text{Ga}^*(\theta, \lambda)$. The moment generating function of Z is then

$$\begin{aligned} Ee^{sZ} &= E\left\{E\left(e^{sZ}|N\right)\right\} \\ &= E\left\{M^*(s; \theta, -\alpha)\right\}^N \\ &= \exp[m\{M^*(s; \theta, -\alpha) - 1\}], \\ &= \exp\left[\lambda\kappa_p(\theta)\left\{\left(1 + \frac{s}{\theta}\right)^\alpha - 1\right\}\right]. \end{aligned}$$

By (4.16), this shows that $Z \sim \text{Tw}_p^*(\theta, \lambda)$. The Tweedie models with $1 < p < 2$ are hence compound Poisson distributions.

We now derive the density of this distribution. The distribution has a positive probability in zero,

$$P(Z = 0) = P(N = 0) = \exp\{-\lambda\kappa_p(\theta)\}. \quad (4.29)$$

SPECIAL CASES

Letting $p^*(z;\theta,\lambda)$ denote the probability density function of the gamma distribution $\mathrm{Ga}^*(\theta,\lambda)$, we find that the joint density of Z and N is, for $n \geq 1$ and $z > 0$,

$$\begin{aligned}
p_{Z,N}(z,n;\theta,\lambda,\alpha) &= p^*(z;\theta,-n\alpha)\frac{m^n}{n!}e^{-m} \\
&= \frac{(-\theta)^{-n\alpha}m^n z^{-n\alpha-1}}{\Gamma(-n\alpha)n!}\exp\{\theta z - m\} \\
&= \frac{\lambda^n \kappa_p^n\left(-\frac{1}{z}\right)}{\Gamma(-n\alpha)\,n!z}\exp\{\theta z - \lambda\kappa_p(\theta)\}.
\end{aligned}$$

The distribution of Z is continuous for $z > 0$, and summing out n in (4.30) the density of Z is

$$p_Z^*(z;\theta,\lambda,\alpha) = \frac{1}{z}\sum_{n=1}^{\infty}\frac{\lambda^n \kappa_p^n\left(-\frac{1}{z}\right)}{\Gamma(-n\alpha)n!}\exp\{\theta z - \lambda\kappa_p(\theta)\}.$$

By the duality transformation, we find that the reproductive version of the Tweedie model $\mathrm{Tw}_p(\mu,\sigma^2)$ has probability density function given by

$$p(y;\theta,\lambda,\alpha) = c_p(y;\lambda)\exp[\lambda\{\theta y - \kappa_p(\theta)\}], \quad y \geq 0, \qquad (4.30)$$

where

$$c_p(y;\lambda) = \begin{cases} \dfrac{1}{y}\sum_{n=1}^{\infty}\dfrac{\lambda^n \kappa_p^n\left(-\frac{1}{\lambda y}\right)}{\Gamma(-\alpha n)n!} & \text{for } y > 0 \\ 1 & \text{for } y = 0. \end{cases} \qquad (4.31)$$

Figure 4.5 shows some compound Poisson density functions for $p = 1.5$.

The reader will notice a certain similarity between the compound Poisson density function (4.30) and the density (4.19) for the stable Tweedie model (Aalen, 1992). In fact, by the reflection formula

$$\Gamma(u)\Gamma(1-u) = \frac{\pi}{\sin(\pi u)},$$

we find that (4.19) may be written in the form

$$c_p^*(z;\lambda) = \frac{1}{z}\sum_{k=1}^{\infty}\frac{\lambda^k \kappa_p^k\left(-\frac{1}{z}\right)}{\Gamma(-\alpha k)k!}, \quad z > 0,$$

which by the duality transformation is analogous to (4.31). The stable case hence corresponds to negative arguments for the gamma function, and in practice we always use the form (4.19).

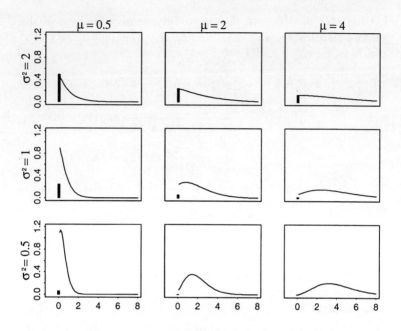

Figure 4.5. *Some compound Poisson Tweedie density functions for* $p = 1.5$. *The probability in zero is indicated by a bar.*

Let us calculate the conditional distribution of N given Z. Define

$$p(z, n; \lambda, \alpha) = \frac{\lambda^n \kappa_p^n \left(-\frac{1}{z}\right)}{\Gamma(-n\alpha)\, n!\, z}.$$

For $z > 0$ we obtain the conditional probability function, for $n = 1, \ldots,$

$$p_{N|Z}(n|z; \lambda, \alpha) = \frac{p(z, n; \lambda, \alpha)}{\sum_{i=1}^{\infty} p(z, i; \lambda, \alpha)}. \quad (4.32)$$

In particular, the conditional mean of N given Z is

$$E(N|Z) = \frac{\sum_{n=1}^{\infty} n p(z, n; \lambda, \alpha)}{\sum_{n=1}^{\infty} p(z, n; \lambda, \alpha)}. \quad (4.33)$$

Note that (4.32) and (4.33) do not involve θ, because Z is sufficient for θ when (λ, α) is known.

Residuals for Tweedie family

Following up on results from Section 3.6, we now consider the definition of residuals for the Tweedie class in the case $p \in (1, 2) \cup$

SPECIAL CASES

$(2, \infty)$, and study the approximation of the distribution function for the compound Poisson case.

For the Tweedie models, the score and Pearson residuals are defined by

$$r_P = s = \frac{y - \mu}{\mu^{\frac{p}{2}}},$$

and the dual score and Wald residuals are defined by

$$r_W = u = \frac{\mu^{1-\frac{p}{2}}}{1-p}\left\{\left(\frac{y}{\mu}\right)^{1-\frac{p}{2}} - \left(\frac{y}{\mu}\right)^{\frac{p}{2}}\right\}.$$

We define the deviance residual by

$$r = \pm\mu^{1-\frac{p}{2}}\sqrt{2\left\{\frac{\left(\frac{y}{\mu}\right)^{2-p}}{(1-p)(2-p)} - \frac{\frac{y}{\mu}}{1-p} + \frac{1}{2-p}\right\}},$$

where $\pm$ denotes the sign of $y - \mu$. Note that, by the scale transformation formula, both r/σ and u/σ are invariant under scale transformations.

Let us consider the case $1 < p < 2$. In this case the above expressions for u and r are also valid for $y = 0$, where r is finite, while $u = 0$. Hence, the modified deviance residual

$$r^* = \frac{r}{\sigma} + \frac{\sigma}{r}\log\frac{u}{r}$$

tends to infinity as y tends to zero, as we saw in the Poisson case. Figure 4.6 shows some quantile plots for r and r^*, with the convention that for $y = 0$ we define

$$r^* = \Phi^{-1}\left\{\frac{1}{2}P(Y = 0)\right\},$$

similar to the Poisson case.

The feature that r^* tends to infinity when y goes to 0 is present in the plots with $\sigma^2 = 1$ and 2. Unfortunately the plots do not show this feature very clearly because of truncation. Otherwise, the plots show that $\Phi(r^*)$ provides an excellent approximation to the distribution function. For intermediate values of σ^2, this approximation is significantly better than the corresponding approximation based on r.

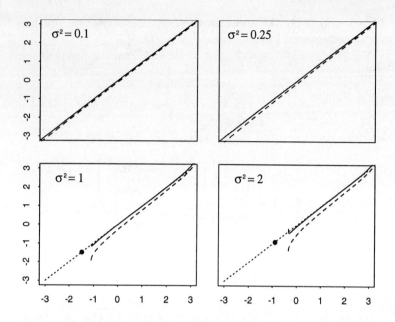

Figure 4.6. *Quantile plots for Tweedie compound Poisson residuals (r^*: solid line, r/σ: - - - -, diagonal: ······, value of r^* at zero: •).*

4.3 Tweedie additive process

4.3.1 General case

Let us consider the additive version of the Tweedie model

$$\mathrm{Tw}_p^*(\theta, \lambda) = \mathrm{Tw}_p\left\{\lambda \tau_p(\theta), \lambda^{1-p}\right\}.$$

The corresponding additive process $Z(t)$, which we call the **Tweedie process**, is defined as follows. If $Z(0) = 0$ and $\Delta Z_i = Z(t_i) - Z(t_{i-1})$ for given time points $0 = t_0 < t_1 < \cdots < t_n$, we define the distribution of the increments of the process by

$$\Delta Z_i \sim \mathrm{Tw}_p\left\{\xi \Delta t_i, (\rho \Delta t_i)^{1-p}\right\},$$

where $\Delta t_i = t_i - t_{i-1}$ and $\xi = \rho \tau_p(\theta)$ denotes the rate of the process. The marginal distribution of the process is then

$$Z(t) \sim \mathrm{Tw}_p(\xi t, t^{1-p}\sigma^2),$$

where $\sigma^2 = \rho^{1-p}$. By the scale transformation property, we obtain

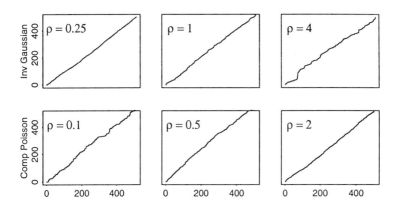

Figure 4.7. *Simulations of the Tweedie additive process, all with rate* $\xi = 1$.

$$\frac{1}{t}Z(t) \sim \mathrm{Tw}_p\left(\xi, \frac{\sigma^2}{t}\right), \tag{4.34}$$

which shows that the empirical rate of the process, $Z(t)/t$, is distributed according to the reproductive version of the model, with t entering as weight.

For a positive distribution, we may interpret this as follows. Let $Z(t)$ be the yield of some process (for example a production process or a biological process) corresponding to the exposure t where t is time, mass, area, etc. By the additive version of the convolution formula, the yield of the process is additive in the exposure t. Formula (4.34) gives the distribution of the average yield per unit of exposure, with mean ξ and dispersion parameter σ^2/t.

The two cases $\xi = \infty$, $p > 2$ and $\xi = 0$, $p < 0$ correspond to stable additive processes, with increments following positive stable and extreme stable distributions, respectively.

The cases $p = 0, 1, 2$ correspond to processes that we have met before, namely the Brownian motion, the Poisson process and the gamma process, respectively. For $p = 3$ we obtain the **inverse Gaussian process**, defined by the following distribution for the increments:

$$\Delta Z_i \sim \mathrm{IG}\left\{\Delta t_i \xi, (\rho \Delta t_i)^{-2}\right\}.$$

This process was studied by Wasan (1968, 1969). Figure 4.7 illustrates the inverse Gaussian and compound Poisson processes.

4.3.2 Compound Poisson process

A special case of the above process is the Tweedie compound Poisson process, which may be defined as follows. Let $N(t)$ be a Poisson process with rate $\lambda \kappa_p(\theta)$, and define $Z(t)$ as 0 if $N(t) = 0$ and otherwise

$$Z(t) = \sum_{i=1}^{N(t)} X_i,$$

where $X_1, X_2, \ldots$ is a sequence of independent $\text{Ga}^*(\theta, -\alpha)$ random variables, independent of $N(t)$. This gives us the Tweedie process, corresponding to

$$Z(t) \sim \text{Tw}_p^*(\theta, \rho t)$$

with $1 < p < 2$.

This process jumps by the amount X_i at the ith event of the Poisson process $N(t)$. The parameter α is the shape parameter for the jump-size distribution; in fact $(-\alpha)^{-1/2}$ is the coefficient of variation. When $|\alpha|$ is large and the coefficient of variation is small, the jumps of the compound Poisson process are nearly of constant size, and so the process is essentially a Poisson process. A large coefficient of variation corresponds to high variability of the jump-size distribution.

4.4 Tweedie convergence results

We now consider a general convergence result for exponential dispersion models, where the Tweedie models appear as limiting distributions. The result applies to models with unit variance functions that behave asymptotically as powers. Our first result shows that variance functions for many common density functions have this asymptotic behaviour.

4.4.1 Regularity of variance functions

Let $\mathcal{P}$ be a natural exponential family with probability mass in zero given by

$$P_\theta(Y = 0) = c(0) \exp\{-\kappa(\theta)\},$$

and assume that the family is continuous for $y > 0$, with probability density function

$$p(y; \theta) = c(y) \exp\{y\theta - \kappa(\theta)\}.$$

The next theorem, from Jørgensen, Martínez and Tsao (1994), concerns the asymptotic behaviour of c and the unit variance function V near zero.

Theorem 4.4 (Jørgensen, Martínez and Tsao) *Suppose that c has the form*

$$c(y) = y^{\gamma-1}g(y) \quad \text{for} \quad y > 0, \tag{4.35}$$

for some $\gamma > 0$, where g is analytic in zero and $g(0) \neq 0$. Then the variance function V of $\mathcal{P}$ satisfies

$$V(\mu) \sim c_0 \mu^p \quad \text{as} \quad \mu \to 0 \tag{4.36}$$

for some $p \in (1, 2]$ and $c_0 > 0$, where

$$p = \begin{cases} \frac{\gamma+2}{\gamma+1} & \text{for} \quad c(0) > 0 \\ 2 & \text{for} \quad c(0) = 0. \end{cases}$$

Proof. Let ν be the measure defined by $\nu(dy) = c(y)\,dy$ for $y > 0$ and $\nu(0) = c(0)$. Define, for $r = 0, 1, \ldots,$

$$M^{(r)}(\theta) = \int_{[0,\infty)} y^r e^{\theta y}\, \nu(dy),$$

where, for any function g,

$$\int_{a_1}^{a_2} g(x)\nu(dx) = \int_{[a_1, a_2)} g(x)\,\nu(dx)$$

for any interval $[a_1, a_2)$ with $a_1 > -\infty$. First note that

$$M^{(r)}(\theta) = c(0)\delta(r) + \int_{(0,\infty)} e^{\theta y} y^{r+\gamma-1} g(y)\, dy$$

where $\delta(0) = 1$ and $\delta(r) = 0$ if $r \geq 1$. Since $\gamma > 0$ and $g(y)$ is analytic at $y = 0$, Watson's Lemma gives the following asymptotic expansion of $M^{(r)}(\theta)$ (see Copson, 1965 and Murray, 1974):

$$M^{(r)}(\theta) \sim c(0)\delta(r) + \sum_{n=0}^{\infty} \frac{g^{(n)}(0)\Gamma(r+\gamma+n)}{n!(-\theta)^{r+\gamma+n}}$$

$$\sim c(0)\delta(r) + \frac{g(0)\Gamma(r+\gamma)}{(-\theta)^{r+\gamma}} \quad \text{as} \quad \theta \to -\infty.$$

If $c(0) > 0$, then

$$\mu = \frac{M^{(1)}(\theta)}{M^{(0)}(\theta)} \sim \frac{g(0)\Gamma(\gamma+1)}{c(0)(-\theta)^{\gamma+1}} \tag{4.37}$$

and
$$V(\mu) = \frac{M^{(2)}(\theta)}{M^{(0)}(\theta)} - \mu^2 \sim \frac{g(0)\Gamma(\gamma+2)}{c(0)(-\theta)^{\gamma+2}}$$
$$\sim c_0 \mu^{\frac{\gamma+2}{\gamma+1}} \quad \text{as} \quad \mu \to 0,$$

where $c_0 > 0$ is a constant. We obtain $p = (\gamma+2)/(\gamma+1) \in (1,2)$, because $\gamma > 0$.

If $c(0) = 0$, then
$$\mu \sim \frac{\Gamma(\gamma+1)}{\Gamma(\gamma)(-\theta)} = \frac{\gamma}{-\theta}$$

and
$$V(\mu) \sim \frac{\Gamma(\gamma+2)}{\Gamma(\gamma)(-\theta)^2} - \mu^2$$
$$\sim \frac{\gamma}{(-\theta)^2}$$
$$\sim \frac{\mu^2}{\gamma}.$$

This proves the theorem. □

Note that, by (4.35), the right limit $c(0^+)$ is infinite for $0 < p < 1$, positive for $p = 1$ and zero for $p > 1$. The corresponding values for p in the case $c(0) > 0$ are $1 < p < 1.5$, $p = 1.5$ and $1.5 < p < 2$, respectively. The asymptotic behaviour of V in (4.36) illustrates the case $\delta = 0$ of Theorem 2.15 (Chapter 2).

4.4.2 Tweedie convergence theorem

We now turn to the Tweedie convergence theorem. Inspired by Theorem 4.4, we define regularity of a unit variance function as follows. Let $\mathrm{ED}(\mu, \sigma^2)$ denote an exponential dispersion model with unit variance function V, mean domain Ω and index set Λ. The unit variance function V is said to be **regular of order** p at zero (respectively at infinity) if
$$V(\mu) \sim c_0 \mu^p,$$
as μ tends to zero (respectively to infinity) for $p \in \mathbf{R}$ and $c_0 > 0$.

Theorem 4.5 (Jørgensen, Martínez and Tsao) *Suppose the unit variance function V is regular of order p at either zero or infinity. Then $p \notin (0,1)$ and for any $\mu > 0$, $\sigma^2 > 0$*
$$c^{-1}\mathrm{ED}(c\mu, \sigma^2 c^{2-p}) \xrightarrow{d} \mathrm{Tw}_p(\mu, c_0\sigma^2) \qquad (4.38)$$

as $c \downarrow 0$ or $c \to \infty$, respectively, where the convergence is through values of c such that $c\mu \in \Omega$ and $c^{p-2}/\sigma^2 \in \Lambda$. The model must be infinitely divisible if $c^{2-p} \to \infty$.

Proof. Consider the case where V is regular at zero (the proof is similar in the case of regularity at infinity). For fixed values of c and σ^2, the left-hand side of (4.38) is a natural exponential family with variance function

$$V_c(\mu) = \sigma^2 c^{-p} V(c\mu) \to \sigma^2 c_0 \mu^p \quad \text{as} \quad c \downarrow 0.$$

If we can show that the convergence is uniform in μ on compact subsets of $\mathbf{R}_+$, the result follows from the Mora convergence theorem. In particular, Mora's convergence theorem implies that the limiting distribution exists, so unless $p \notin (0,1)$ we would have a contradiction with Theorem 4.2.

To show that the convergence is uniform in μ on compact subsets of $\mathbf{R}_+$, let $0 < \mu \le M$ and let μ_0 be such that

$$\left| \frac{V(\mu)}{\mu^p} - c_0 \right| < \varepsilon$$

for all $0 < \mu < \mu_0$. Then for any $c < \mu_0/M$

$$\left| \frac{V(c\mu)}{c^p} - c_0 \mu^p \right| = \mu^p \left| \frac{V(c\mu)}{(c\mu)^p} - c_0 \right| \le M^p \varepsilon,$$

which shows the convergence to be uniform, concluding the proof. □

This is a theorem with wide implications because all the unit variance functions that we have met so far are regular at either zero or infinity, and we have already met several special cases of the theorem in Chapters 2 and 3. The theorem furthermore suggests that a wide range of exponential dispersion models may be approximated by Tweedie models.

Types of convergence

We distinguish between the following three types of convergence in Theorem 4.5. The **central limit type** is defined as the case where the dispersion parameter $\sigma^2 c^{2-p}$ goes to zero. The **infinitely divisible type** is defined as the case where the dispersion parameter goes to infinity, and the **Tauber type** is defined as the case where the dispersion parameter is constant, meaning regularity of order $p = 2$. When Theorem 4.5 applies to an exponential dispersion model $\mathrm{ED}(\mu, \sigma^2)$, we say that $\mathrm{ED}(\mu, \sigma^2)$ **is locally** $\mathrm{Tw}_p(\mu, \sigma^2)$ **at zero**, respectively **at infinity**.

The central limit type corresponds to regularity at zero with orders $p < 2$ or at infinity with orders $p > 2$. This case may be considered an extension of classical limit theory for sums of independent and identically distributed random variables, in the following sense. By the additive version of the convolution formula, we find that a small value of the dispersion parameter, which corresponds to a large value of the index parameter, may be interpreted as a variable that is a sum of a large number of independent and identically distributed variables. The difference with classical theory is that the limiting result (4.38) requires the mean $c\mu$ to go to either zero or infinity, whereas conventional central limit theory holds for a fixed distribution of the components of the sum.

The infinitely divisible type includes regularity at zero with orders $p > 2$ and regularity at infinity with orders $p < 2$. By contrast with the central limit type, the infinitely divisible type requires the distribution to be infinitely divisible, because the dispersion parameter $\sigma^2 c^{2-p}$ goes to infinity, which is only possible if $\Lambda = \mathbf{R}_+$. Note in particular that if $\Lambda = \mathbf{N}$, then the dispersion parameter takes values in the set $\{1, \frac{1}{2}, \frac{1}{3}, \ldots\}$, which is bounded, so that $\sigma^2 c^{2-p}$ cannot go to infinity through values in Λ^{-1}. This appears to be a new form of convergence, dealing with the limiting distribution of a random variable 'divided' by n for n going to infinity, in the sense of division as defined in Chapter 3.

Finally, the terminology 'Tauber type' is due to its interpretation in terms of Tauber theory. The reader is referred to Feller (1971, p. 445) for details about Tauber theory. The Tauber type corresponds to the case $p = 2$ in Theorem 4.4, giving rise to convergence to the gamma distribution. This type of convergence is different from the two other types just defined in the sense that the dispersion parameter $\sigma^2 c^{2-p}$ is fixed in this case, and the convergence is obtained for the mean $c\mu$ going either to zero or to infinity. We consider this case in some more detail below.

The Tweedie convergence theorem is probably related with the stable generalization of the Central Limit Theorem (Feller, 1971, pp. 574–581). The following seems to be a reasonable conjecture. If a distribution belongs to the domain of attraction of a positive stable or extreme stable distribution, then probably the exponential dispersion model generated from it converges according to the Tweedie convergence theorem. One reason that such a result is not as straightforward to prove as the Tweedie convergence theorem itself it that it involves values of μ at the extremes of Ω, where Mora's convergence theorem, as stated, does not apply.

4.4.3 Convergence of Tauber type

Before stating the main result of this section, we consider some basic facts about regular variation. A function $L : \mathbf{R}_+ \to \mathbf{R}_+$ is said to **vary slowly at infinity** if, for all $x > 0$,

$$\lim_{t \to \infty} \frac{L(tx)}{L(t)} = 1. \tag{4.39}$$

The function L is said to **vary slowly at zero** if $L(1/x)$ varies slowly at infinity. A function $u : \mathbf{R}_+ \to \mathbf{R}_+$ is said to **vary regularly at infinity (respectively zero) with exponent** $\gamma \in \mathbf{R}$, if there exists a slowly varying function L, such that

$$u(x) = x^\gamma L(x). \tag{4.40}$$

A measure ν on $\mathbf{R}_+$ is said to vary regularly with exponent γ if the improper distribution function

$$\bar{\nu}(x) = \nu\{(0, x]\}$$

varies regularly with exponent γ at the appropriate extreme of $\mathbf{R}_+$. In particular, if ν has density with respect to Lebesgue measure of the form

$$\nu(dx) = g(x) x^{\gamma - 1}, \tag{4.41}$$

where g satisfies the conditions of Theorem 4.4, then ν is regularly varying at zero with exponent γ. As we saw in Theorem 4.4, this implies that the unit variance function of the natural exponential family generated by ν is regular at zero of order 2, which in turn implies that the family is locally gamma at zero.

The classical Tauber theorem (see Feller, 1971, p. 445) relates regular variation of a measure to the asymptotic behaviour of its moment generating function, see also de Haan (1975) and Bingham, Goldie and Teugels (1987).

Theorem 4.6 (Tauber Theorem) *Let ν be a measure on $\mathbf{R}_+$ such that its Laplace transform*

$$\omega(t) = \int_0^\infty e^{-tx} \nu(dx)$$

exists for $t > 0$. Let L be slowly varying at infinity (respectively *zero) and let* $0 \leq \gamma < \infty$. *Then*

$$\bar{\nu}(t) \sim \frac{1}{\Gamma(\gamma+1)} t^\gamma L(t) \Leftrightarrow \omega\left(\frac{1}{t}\right) \sim t^\gamma L(t)$$

as $t \to \infty$ *(respectively* $t \to 0$*)*.

If we consider the measure ν of the form (4.41), where g satisfies the conditions of Theorem 4.4, then the Tauber Theorem implies that the cumulant function of the natural exponential family generated by ν has the form

$$e^{\kappa(\theta)} = (-\theta)^{-\gamma} L(-\theta),$$

where L is slowly varying at infinity. The next result, due to Jørgensen, Martínez and Tsao (1994), shows that a cumulant function of this form implies gamma convergence under much weaker conditions and extends it to regular variation at infinity.

Theorem 4.7 (Jørgensen, Martínez and Tsao) *Let* $\mathrm{ED}(\mu, \sigma^2)$ *be generated by the measure* ν *with support* $S \subseteq (0, \infty)$. *Suppose that* ν *is regularly varying at either zero or infinity, so that by the Tauber Theorem, the moment generating function for* ν *has the form*

$$M_\nu(\theta) = (-\theta)^{-\gamma} L(-\theta), \quad \theta < 0, \tag{4.42}$$

for some $\gamma > 0$, *where* L *is slowly varying at either infinity or zero, respectively. If* $f(x) = \log L(x)$ *satisfies*

$$\lim_{x \to \infty} x f'(x) = 0, \tag{4.43}$$

or

$$\lim_{x \to 0} f'(x) = 0,$$

respectively, then for any $\mu > 0$ *and any* σ^2 *such that* $1/\sigma^2 \in \Lambda$

$$\frac{1}{c} \mathrm{ED}(c\mu, \sigma^2) \xrightarrow{d} \mathrm{Ga}(\mu, \sigma^2/\gamma) \tag{4.44}$$

for c *tending to zero or infinity, respectively.*

Proof. Consider the case where ν is regularly varying at zero, so that L is slowly varying at infinity. The mean value mapping for the natural exponential family generated by ν is

$$\tau(\theta) = \frac{\gamma}{-\theta} - f'(-\theta), \quad \theta < 0. \tag{4.45}$$

By (4.43) we have $\lim_{\theta \to -\infty} f'(-\theta) = 0$, and hence $\lim_{\theta \to -\infty} \tau(\theta) = 0$, or $\inf \Omega = 0$. Inserting $\theta = \tau^{-1}(\mu)$ in (4.45) we obtain

$$\tau^{-1}(\mu) = \frac{-\gamma}{\mu + f'\{-\tau^{-1}(\mu)\}}, \quad \mu \in \Omega. \tag{4.46}$$

The reproductive model $ED(\mu, \sigma^2)$ generated by ν has moment generating function of the following form, for $\mu = \tau(\theta)$, $\sigma^2 = 1/\lambda$ and $s < -\lambda\theta$:

$$M(s;\theta,\lambda) = \frac{M_\nu^\lambda(\theta + s/\lambda)}{M_\nu^\lambda(\theta)} = \left(1 + \frac{s}{\lambda\theta}\right)^{-\lambda\gamma} \frac{L^\lambda(-\theta - s/\lambda)}{L^\lambda(-\theta)}.$$

The moment generating function for the left-hand side of (4.44) is hence, for $\mu > 0$ fixed and c small enough to make $c\mu \in \Omega$, given by

$$M\left\{\frac{s}{c}; \tau^{-1}(c\mu)\right\} = \left\{1 + \frac{s}{\lambda c \tau^{-1}(c\mu)}\right\}^{-\lambda\gamma} h^\lambda(s; c, \mu, \lambda)$$

where

$$h(s; c, \mu, \lambda) = \frac{L\left\{-\tau^{-1}(c\mu) - \frac{s}{\lambda c}\right\}}{L\left\{-\tau^{-1}(c\mu)\right\}}.$$

If we can show

$$\lim_{c \to 0} c\tau^{-1}(c\mu) = \frac{-\gamma}{\mu} \qquad (4.47)$$

and

$$\lim_{c \to 0} h(s; c, \mu, \lambda) = 1, \qquad (4.48)$$

it follows that

$$\lim_{c \to 0} M\left\{\frac{s}{c}; \tau^{-1}(c\mu)\right\} = \left(1 - \frac{\mu s}{\lambda \gamma}\right)^{-\lambda\gamma},$$

which by the continuity theorem for moment generating functions implies (4.44).

To show (4.47) and (4.48), we first show that, for $\mu > 0$ fixed,

$$\lim_{c \to 0} \frac{1}{c} f'\left\{-\tau^{-1}(c\mu)\right\} = 0. \qquad (4.49)$$

Let $\delta = \tau^{-1}(c\mu)$ so that $c \to 0$ is equivalent to $\delta \to -\infty$, for μ fixed. Inserting $c = \tau(\delta)/\mu$ in (4.49) we obtain

$$\lim_{\delta \to -\infty} \frac{\mu}{\tau(\delta)} f'(-\delta) = \lim_{\delta \to -\infty} \frac{\mu}{\tau(\delta)(-\delta)}(-\delta) f'(-\delta),$$

so by (4.43) it is enough to show that the limit of $-\delta\tau(\delta)$ is finite. From (4.45) with $\theta = \delta$ we obtain, using (4.43),

$$\lim_{\delta \to -\infty} \{-\delta\tau(\delta)\} = \lim_{\delta \to -\infty} \{\gamma + \delta f'(-\delta)\} = \gamma \qquad (4.50)$$

so we have shown (4.49). Now if we take $\mu = \mu c$ in (4.46) we obtain

$$c\tau^{-1}(c\mu) = \frac{-\gamma}{\mu + \frac{1}{c} f'\left\{-\tau^{-1}(c\mu)\right\}},$$

so by (4.49) we have shown (4.47).

To show (4.48) we take $\delta = \tau^{-1}(c\mu)$ in the definition of h, obtaining

$$h(s; c, \mu, \lambda) = \frac{L\left[-\delta\left\{1 + \frac{s\mu}{\lambda\delta\tau(\delta)}\right\}\right]}{L(-\delta)}. \quad (4.51)$$

By Karamata's Theorem (see e.g. de Haan, 1975, p. 19) there exist functions $\ell(x)$ and $a(x)$ satisfying

$$\lim_{x \to \infty} \ell(x) = c_0 \quad (0 < c_0 < \infty) \quad (4.52)$$

and

$$\lim_{x \to \infty} a(x) = 0, \quad (4.53)$$

such that for all $x > 0$

$$L(x) = \ell(x) \exp\left\{\int_1^x \frac{a(t)}{t} dt\right\}. \quad (4.54)$$

Hence, taking $u(\delta) = s\mu/\{\lambda\delta\tau(\delta)\}$, (4.51) becomes

$$\frac{L\left[-\delta\{1 + u(\delta)\}\right]}{L(-\delta)} = \frac{\ell\left[-\delta\{1 + u(\delta)\}\right]}{\ell(-\delta)} \exp\left[\int_{-\delta}^{-\delta\{1+u(\delta)\}} \frac{a(t)}{t} dt\right]. \quad (4.55)$$

By (4.50), $\lim_{\delta \to -\infty} u(\delta) = -s\mu/(\lambda\gamma)$, so the first factor of (4.55) tends to 1, and for $s < 0$ we have $u(\delta) > 0$, in which case

$$\int_{-\delta}^{-\delta\{1+u(\delta)\}} \frac{a(t)}{t} dt \le u(\delta) \sup_{-\delta < t < -\delta\{1+u(\delta)\}} a(t),$$

which tends to zero, so we have shown (4.48), and hence (4.44). The proof in the case where ν is regularly varying at infinity is analogous, and we omit the details. $\square$

By using Karamata's Theorem, formula (4.54), we find that (4.43) is equivalent to

$$\lim_{x \to \infty} \left\{x \frac{\ell'(x)}{\ell(x)} + a(x)\right\} = 0,$$

which by (4.52) is, in turn, equivalent to

$$\lim_{x \to \infty} x\ell'(x) = 0. \quad (4.56)$$

Since any pair of functions ℓ and a satisfying (4.52) and (4.53) define a slowly varying function via (4.54), the condition (4.56) cannot be derived from the fact that L is slowly varying alone.

However, the theorem may possibly be sharpened by taking into account that $L(x)$ is the ratio of the moment generating function M_ν and $(-\theta)^{-\gamma}$.

A variance function that is regular of order γ in the sense defined in Section 4.4.2 is also regularly varying with exponent γ. However, only the value $\gamma = 2$ is associated with regular variation of the generating measure ν. For example, in the case $\gamma = 3$, the inverse Gaussian density is not regularly varying at zero. Some results about regular variation of the Levy measure of an infinitely divisible exponential dispersion model have been obtained by Jørgensen and Martínez (1997).

4.4.4 Examples

Example 4.1 The inverse binomial distribution of Yanagimoto (1989) is defined by the probability function for $z \in \mathbf{N}_0$,

$$p(z; \rho, \lambda) = \frac{\lambda \Gamma(2x + \lambda)}{\Gamma(z+1)\Gamma(z+\lambda+1)} \{\rho(1-\rho)\}^z \rho^\lambda, \qquad (4.57)$$

where $\lambda > 0$ and $\frac{1}{2} < \rho < 1$. This model is an additive family with cubic unit variance function

$$V(\mu) = \mu(\mu+1)(2\mu+1)$$

for $\mu > 0$. This unit variance function is regular of order 1 at zero and order 3 at infinity. It is hence locally Poisson at zero, and locally inverse Gaussian at infinity, as shown by Yanagimoto (1989) by direct methods. Both convergences are of central limit type. The inverse binomial distribution is a special case of the generalized negative binomial distribution of Jain and Consul (1971). □

Example 4.2 The generalized Poisson distribution (Consul and Jain, 1973) is defined for $z \in \mathbf{N}_0$ by the probability function

$$p(z; \psi, \rho) = \frac{\rho(\rho + \psi z)^{z-1}}{z!} \exp\{-(\rho + \psi z)\}.$$

Let us take the parameter domain to be $\rho > 0$, $0 < \psi < 1$, where the limiting case $\psi = 0$ gives the Poisson distribution. The reparametrization from ρ to $\lambda = \rho/\psi > 0$ gives the following form of the probability function:

$$p(z; \psi, \rho) = \frac{\lambda(\lambda + z)^{z-1}}{z!} \exp\{z(\log\psi - \psi) - \lambda\psi\}, \qquad (4.58)$$

which shows that we have an additive family with canonical

parameter $\theta = \log \psi - \psi$. The canonical parameter domain is $\Theta = (-\infty, -1)$. The unit cumulant function $\kappa(\theta)$ is given as the solution of the equation $\log \psi - \psi = \theta$, and does not have a closed-form expression. However, the mean is $\mu = \lambda \psi/(1-\psi)$, and the unit variance function is cubic,

$$V(\mu) = \mu(1+\mu)^2, \quad \mu > 0,$$

for details see Exercise 4.12. Again this unit variance function is regular of order 1 at zero and order 3 at infinity, and the family is locally Poisson at zero and locally inverse Gaussian at infinity. In both cases the convergence is of central limit type. □

We note that the unit variance functions of both the above examples are cubic polynomials. Letac and Mora (1990) studied cubic variance functions and found six types of strictly cubic variance functions, besides the six quadratic ones found by Morris (see Section 3.4). Letac and Mora (1990) referred to the classes containing (4.57) and (4.58) as the 'Takács' and 'Abel' classes, respectively. All strictly cubic variance functions are regular of order 3 at infinity; for details see Exercise 4.15.

Example 4.3 Consider the reproductive model generated by the half-normal distribution with density function

$$p(x) = \sqrt{\frac{2}{\pi}} e^{-\frac{1}{2}x^2}, \qquad x > 0. \tag{4.59}$$

This density is of the form required in Theorem 4.4 with $\gamma = 1$ and zero probability at zero. It follows that the corresponding unit variance function is regular of order 2 at zero. Hence, by the Tweedie convergence theorem, the model is locally gamma at zero. □

Example 4.4 (Letac, 1992, pp. 100–101.) The measure ν defined by

$$\nu(dy) = (e^{2y} - 1)\, dy, \qquad y > 0,$$

generates a natural exponential family $\mathrm{NE}(\mu)$ with the following variance function (Exercise 4.16):

$$V(\mu) = \mu^2 + 1 - \sqrt{\mu^2 + 1}, \quad \mu > 0. \tag{4.60}$$

This variance function is of order 2 at zero as well as at infinity. Hence, by Theorem 4.5 we obtain the following convergence results, both of Tauber type:

$$\frac{1}{c}\mathrm{NE}(c\mu) \xrightarrow{d} \mathrm{Ex}(\mu) \quad \text{as} \quad c \to \infty$$

and
$$\frac{1}{c}\mathrm{NE}(c\mu) \xrightarrow{d} \mathrm{Ga}\left(\mu, \frac{1}{2}\right) \quad \text{as} \quad c \to 0.$$
□

The variance function (4.60) is an example of the form of variance function proposed by Letac (1987, 1992). The form considered by Letac (1992) is

$$V(\mu) = P(\mu)R(\mu) + Q(\mu)\sqrt{R(\mu)}, \tag{4.61}$$

where P, Q and R are polynomials. We say that such a variance function is of **Letac form**. A special case of this class of variance functions was studied by Kokonendji (1994). Variance functions of Letac form are regular of integer or half-integer orders.

Example 4.5 (Letac, 1992, p. 96.) Let us define the measure ν as follows:
$$\nu(dy) = \delta_0(dy) + 1_{(0,\infty)}(y)\, dy,$$
where δ_n denotes the Dirac-delta measure in n. The exponential dispersion model generated by ν has variance function of Letac form,
$$V(\mu) = \mu\sqrt{\mu^2 + 4\mu}, \quad \mu > 0.$$
This variance function is regular of order 1.5 at zero and of order 2 at infinity. The family is hence locally compound Poisson at zero, the convergence being of central limit type, and locally gamma at infinity. □

Example 4.6 (Letac, 1992, p. 84.) Consider the exponential dispersion model generated by the measure
$$\nu(dz) = \sum_{n=0}^{\infty} a_n \delta_n(dz)$$
where the a_ns are the coefficients of the series
$$\exp\left(\frac{2\lambda x}{1-x}\right) = \sum_{n=0}^{\infty} a_n x^n. \tag{4.62}$$
Since any $\lambda > 0$ is allowed in (4.62), we obtain $\Lambda = \mathbf{R}_+$, so the model is infinitely divisible. The variance function is of Letac form,
$$V(\mu) = \mu\sqrt{1 + 2\mu}, \quad \mu > 0.$$
It is regular of order 1 at zero and order 1.5 at infinity, so the model is locally Poisson at zero, and locally compound Poisson at infinity. The latter convergence is of infinitely divisible type. □

Example 4.7 (Letac, 1992, p. 74.) Let X_1 be distributed according to the positive stable distribution with index $1/2$, given by the density
$$p(y) = \frac{1}{\sqrt{2\pi y^3}} \exp\left(-\frac{1}{2y}\right), \quad y > 0,$$
and let X_2 be independent of X_1, with exponential distribution with parameter 1. Consider the exponential dispersion model generated by the distribution of $Y = X_1 X_2$. The variance function is of Letac form,
$$V(\mu) = \mu^3 + 2\mu^2 + \mu^2\sqrt{\mu^2 + 2\mu}, \quad \mu > 0.$$
This variance function is of order 2 at zero and order 3 at infinity. The model is hence locally gamma at zero, and locally inverse Gaussian at infinity, the latter convergence being of central limit type. □

Generalized inverse Gaussian distribution

The generalized inverse Gaussian distribution $\mathrm{GIG}(\gamma, \chi, \psi)$ (Barndorff-Nielsen, 1977, 1978b; Jørgensen, 1982) is useful for illustrating the many different possible limiting behaviours of the Tweedie convergence theorem. Its probability density function is
$$p(y; \gamma, \chi, \psi) = \frac{(\psi/\chi)^{\gamma/2}}{2K_\gamma(\sqrt{\chi\psi})} y^{\gamma-1} \exp\left\{-\frac{1}{2}(\chi y^{-1} + \psi y)\right\} \quad (4.63)$$
for $y > 0$, where K_γ denotes the modified Bessel function of the third kind with index $\gamma \in \mathbf{R}$. The inverse Gaussian distribution is obtained as the special case where $\gamma = -1/2$. The parameter domain for (χ, ψ) is $\mathbf{R}_+^2$ for $\gamma = 0$, $\mathbf{R}_0 \times \mathbf{R}_+$ for $\gamma > 0$ and $\mathbf{R}_+ \times \mathbf{R}_0$ for $\gamma < 0$. In the cases $\chi = 0$ and $\psi = 0$, which correspond to the gamma and reciprocal gamma distributions, respectively, the normalizing constant in (4.63) should be interpreted in the limiting sense.

Note first that (4.63) is a natural exponential family for each fixed value of (γ, χ) with canonical parameter $\theta = \psi/2$ and unit cumulant function
$$\kappa_{\gamma,\chi}(\theta) = \gamma \log \eta + \log 2 + \log K_\gamma(\omega), \quad (4.64)$$
where $\eta = \sqrt{\chi/\psi}$ and $\omega = \sqrt{\chi\psi}$ are considered as functions of θ. The expectation is
$$\mu = \tau_{\gamma,\chi}(\theta) = \eta \frac{K_{\gamma+1}(\omega)}{K_\gamma(\omega)},$$

and the mean domain is

$$\Omega_{\gamma,\chi} = \begin{cases} \mathbf{R}_+ & \text{for } \gamma \in [-1, 0) \\ (0, b) & \text{for } \gamma < -1, \end{cases}$$

where

$$b = \frac{-\chi}{2(\gamma + 1)}.$$

There is no closed-form expression for the variance function, but it may be expressed as follows in terms of $\tau_{\gamma,\chi}^{-1}$:

$$V_{\gamma,\chi}(\mu) = \frac{2(\gamma+1)\mu + \chi}{-2\tau_{\gamma,\chi}^{-1}(\mu)} - \mu^2, \tag{4.65}$$

for $\mu \in \Omega_{\gamma,\chi}$. We now show that this variance function is regular at the upper terminal of $\Omega_{\gamma,\chi}$, by using asymptotic results for K_γ from Jørgensen (1982, pp. 14 and 170–173).

Let $\text{ED}_{\gamma,\chi}(\mu, \sigma^2)$ denote the reproductive exponential dispersion model generated by the generalized inverse Gaussian distribution, corresponding to a density of the form

$$p(y; \theta, \lambda, \gamma, \chi) = c_{\gamma,\chi}(y; \lambda) \exp\left[\lambda \left\{\theta y - \kappa_{\gamma,\chi}(\theta)\right\}\right].$$

The generalized inverse Gaussian distribution is infinitely divisible, so the domain for λ is $\mathbf{R}_+$. The family has unit variance function (4.65). The following convergence results were shown by Jørgensen, Martínez and Tsao (1994).

Let us define

$$p = \frac{\gamma + 2}{\gamma + 1}.$$

Then the asymptotic behaviour of $V_{\gamma,\chi}$ at the right-hand end point of $\Omega_{\gamma,\chi}$ is as follows (c_1, c_2 and c_3 denote constants).

1. For $\gamma \in (-1, 0)$ we obtain $p > 2$ and

$$V_{\gamma,\chi}(\mu) \sim c_1 \mu^p \quad \text{as} \quad \mu \to \infty,$$

which implies

$$c^{-1}\text{ED}_{\gamma,\chi}(c\mu, c^{2-p}\sigma^2) \xrightarrow{d} \text{Tw}_p(\mu, c_1\sigma^2) \quad \text{as} \quad c \to \infty.$$

2. For $p = -1$ we obtain

$$V_{\gamma,\chi}(\mu) \sim \chi^2 \exp\left(\frac{2\mu}{\chi}\right) \quad \text{as} \quad \mu \to \infty.$$

This case will be considered in Example 4.8 below.

3. For $\gamma \in (-2, -1)$ we obtain $p < 0$ and
$$V_{\gamma,\chi}(\mu) \sim c_2 (b - \mu)^p \quad \text{as} \quad \mu \to b,$$
which implies
$$c^{-1} \left\{ b - \text{ED}_{\gamma,\chi}(b - c\mu, c^{2-p}\sigma^2) \right\} \xrightarrow{d} \text{Tw}_p(\mu, c_2\sigma^2) \quad \text{as} \quad c \to 0.$$

4. For $\gamma < -2$ we obtain
$$V_{\gamma,\chi}(\mu) \sim c_3 \quad \text{as} \quad \mu \to b,$$
which implies
$$c^{-1} \left\{ b - \text{ED}_{\gamma,\chi}(b - c\mu, c^2\sigma^2) \right\} \xrightarrow{d} N(\mu, c_3\sigma^2) \quad \text{as} \quad c \to 0.$$

In all cases, the convergence is of central limit type. The generalized inverse Gaussian distribution hence provides examples of unit variance functions that are regular with orders $p > 2$ and $p \leq 0$.

4.5 Exponential variance functions

4.5.1 Translation of exponential dispersion models

As we saw in Theorem 4.1, the class of Tweedie models is characterized as the class of exponential dispersion models closed under scale transformations. We now consider the same question for translations. For example, we know that the normal distribution is closed under translations,
$$c + N(\mu, \sigma^2) = N(c + \mu, \sigma^2),$$
for $c \in \mathbf{R}$. We may hence ask whether there exist other exponential dispersion models that are closed under translations. The next result shows that such models correspond to exponential unit variance functions.

Theorem 4.8 *Let $ED(\mu, \sigma^2)$ denote a reproductive exponential dispersion model closed under translation, in the sense that there exists a function $f(c, \sigma^2)$ such that*
$$c + \text{ED}(\mu, \sigma^2) = \text{ED}\left\{ c + \mu, f(c, \sigma^2) \right\} \tag{4.66}$$
for any $\mu \in \Omega$, $\sigma^{-2} \in \Lambda$ and $c \in \mathbf{R}$. Then the model is infinitely divisible and has exponential unit variance function.

Proof. By letting c vary in $\mathbf{R}$ we find from the right-hand side of (4.66) that $\Omega = \mathbf{R}$. Calculating the variances on both sides of

(4.66) and taking $\sigma^2 = 1$ gives
$$V(c + \mu) = g(c)V(\mu), \qquad (4.67)$$
for $g(c) = 1/f(c, 1)$. Since V is differentiable and positive, the same is the case for g, and $g(0) = 1$. Differentiating (4.67) with respect to c and letting $c = 0$, we find the differential equation
$$V'(\mu) = g'(0)V(\mu), \qquad (4.68)$$
whose solution is
$$V(\mu) = c_0 \exp\{g'(0)\mu\}$$
where $c_0 > 0$ is a constant. Calculating again the variances on both sides of (4.66) we obtain $f(c, \sigma^2) = \sigma^2 \exp\{-g'(0)c\}$. If $g'(0) \neq 0$, the infinite divisibility follows by letting c vary in $\mathbf{R}$. In the case $g'(0) = 0$, the unit variance function is constant, corresponding to the normal distribution, which is infinitely divisible. $\square$

Absorbing the constant c_0 into σ^2 and defining the new parameter $\beta = g'(0)$, we obtain the unit variance function
$$V(\mu) = e^{\beta\mu}. \qquad (4.69)$$
Let
$$\mathrm{Tw}_\infty(\mu, \sigma^2, \beta)$$
denote the reproductive exponential dispersion model corresponding to (4.69), which we prove the existence of in Section 4.5.3. Figure 4.8 shows some unit deviances for this model.

The corresponding version of (4.66) is
$$c + \mathrm{Tw}_\infty(\mu, \sigma^2, \beta) = \mathrm{Tw}_\infty(c + \mu, \sigma^2 e^{-\beta c}, \beta). \qquad (4.70)$$
The family is also closed under scale transformations, that is, for $c \neq 0$ we find
$$c\mathrm{Tw}_\infty(\mu, \sigma^2, \beta) = \mathrm{Tw}_\infty\left(c\mu, \sigma^2 c^2, \beta/c\right). \qquad (4.71)$$
This formula may be verified by calculating the unit variance function on both sides of the equation. Taking $c = \lambda = \sigma^{-2}$ in this formula gives the duality transformation; hence the additive form of the model is
$$\mathrm{Tw}^*_\infty(\theta, \lambda, \beta) = \mathrm{Tw}_\infty\left\{\lambda\tau_\beta(\theta), \lambda, \frac{\beta}{\lambda}\right\}.$$
Taking $c = -\mu$ in (4.70) and in turn using (4.71) gives
$$\begin{aligned}\mathrm{Tw}_\infty(\mu, \sigma^2, \beta) &= \mu + \mathrm{Tw}_\infty(0, \sigma^2 e^{\beta\mu}, \beta) \qquad (4.72)\\ &= \mu + \beta^{-1}\mathrm{Tw}_\infty(0, \sigma^2 \beta^2 e^{\beta\mu}, 1).\end{aligned}$$

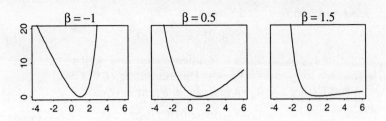

Figure 4.8. *Some unit deviances for the Tweedie model with $p = \infty$.*

This shows that the model is a location-scale model for fixed value of the shape parameter $\sigma^2 \beta^2 e^{\beta\mu}$, with location μ and scale $1/\beta$.

By a location and scale change, the Tweedie unit variance function μ^p may be put in the form

$$V(\mu) = \left(1 + \frac{\beta\mu}{p}\right)^p \to e^{\beta\mu} \quad \text{as} \quad p \to \infty, \qquad (4.73)$$

so that the exponential unit variance function may be considered a limiting case of the power unit variance functions.

4.5.2 Convolution formula

Because of the location-scale form of the distribution, we may prove a more general convolution result than the standard result for exponential dispersion models. Thus, assume that $Y_1, \ldots, Y_n$ are independent, with distribution

$$Y_i \sim \text{Tw}_\infty\left(\mu_i, \frac{\sigma^2}{w_i}, \beta\right) = \mu_i + \text{Tw}_\infty\left(0, \frac{\sigma^2 e^{\beta\mu_i}}{w_i}, \beta\right), \qquad (4.74)$$

where we have used (4.72). Applying the reproductive version of the convolution formula to the second term on the right-hand side of (4.74) and using (4.72) again we obtain

$$\frac{1}{w_+} \sum_{i=1}^n w_i \exp\left(-\beta\mu_i\right) Y_i \sim \text{Tw}_\infty\left(\bar{\mu}, \frac{\sigma^2}{w_+} e^{-\beta\bar{\mu}}, \beta\right) \qquad (4.75)$$

where

$$\bar{\mu} = \frac{1}{w_+} \sum_{i=1}^n w_i \mu_i \exp\left(-\beta\mu_i\right)$$

and
$$w_+ = \sum_{i=1}^{n} w_i \exp(-\beta \mu_i).$$

For $\beta = 0$, formula (4.75) becomes the usual convolution formula for the normal distribution.

4.5.3 Stable distribution with index 1

In Section 4.2.1 we showed that extreme stable distributions with stability index $0 < \alpha < 1$ or $1 < \alpha \le 2$ generate Tweedie models. We now show that exponential dispersion models with exponential unit variance functions are generated from the extreme stable distribution with index $\alpha = 1$. Note that the Cauchy distribution is stable with index $\alpha = 1$, but whereas the Cauchy is a symmetric distribution, the extreme stable distribution with index $\alpha = 1$ is not symmetric.

Due to (4.72) it is enough to study the case $\beta = 1$. Hence, consider the unit variance function
$$V(\mu) = e^\mu \tag{4.76}$$
defined on $\Omega = \mathbf{R}$. Going through the familiar steps of the process leading from V to the unit cumulant function, we first solve the equation
$$\frac{\partial \tau^{-1}}{\partial \mu} = e^{-\mu},$$
with solution $-e^{-\mu}$, giving
$$\tau(\theta) = -\log(-\theta)$$
for $\theta < 0$. This, in turn, leads to
$$\kappa(\theta) = \theta \{1 - \log(-\theta)\},$$
which is defined on $\Theta = (-\infty, 0]$; in particular, $\kappa(0) = 0$ by continuity.

Lemma 4.9 *The function $\kappa(\theta) = \theta \{1 - \log(-\theta)\}$ is the cumulant generating function of a stable distribution with $\alpha = 1$.*

Proof. We want to show that if κ is a cumulant generating function, then the corresponding distribution is stable with $\alpha = 1$. Suppose that $Y_1, \ldots, Y_n$ are independent and identically distributed with cumulant generating function $\kappa(s)$. Then $Y_+ = Y_1 + \cdots + Y_n$ has cumulant generating function

$$\begin{aligned}n\kappa(s) &= ns\left\{1 - \log(-s)\right\}\\ &= ns\left\{1 - \log(-ns)\right\} + ns\log n\\ &= \kappa(ns) + sn\log n.\end{aligned}$$

This shows that the distribution of Y_+ is the same as that of $nY_1 + n\log n$, or

$$n^{-1}Y_+ - \log n \sim Y_1.$$

Hence we have a stable distribution with index $\alpha = 1$. This proves that κ is the cumulant generating function of an extreme stable distribution, see also Eaton, Morris and Rubin (1971). □

The lemma shows that the distribution $\mathrm{Tw}_\infty(\mu, \sigma^2, \beta)$ is the exponential dispersion model generated by an extreme stable distribution with index $\alpha = 1$. Since such distributions are continuous and have support $\mathbf{R}$, the same is the case for $\mathrm{Tw}_\infty(\mu, \sigma^2, \beta)$. Note that the value $\alpha = 1$ gives $p = \infty$ in equation (4.9), which explains the notation Tw_∞. We call this the **Tweedie model with $p = \infty$**.

4.5.4 Convergence results

We have seen that the Tweedie model $\mathrm{Tw}_\infty(\mu, \sigma^2, \beta)$ has many analogies with the other Tweedie models. The following result shows a parallel to the convergence results of Theorem 4.5.

Let $\mathrm{ED}(\mu, \sigma^2)$ denote an exponential dispersion model with unit variance function V, mean domain Ω and index set Λ. The unit variance function V is said to be **regular of order ∞** at plus or minus infinity with coefficient $\beta \in \mathbf{R}$ if, for some $c_0 > 0$,

$$V(\mu) \sim c_0 e^{\beta\mu},$$

as μ tends to plus or minus infinity, respectively.

Theorem 4.10 (Jørgensen, Martínez and Tsao) *Suppose the unit variance function V is regular of order ∞ at either plus or minus infinity with coefficient β. Then for any $\mu \in \mathbf{R}$ and $\sigma^2 > 0$*

$$-c + \mathrm{ED}\left(\mu + c, \sigma^2 e^{-\beta c}\right) \xrightarrow{d} \mathrm{Tw}_\infty(\mu, c_0\sigma^2, \beta), \qquad (4.77)$$

as $c \to \pm\infty$, respectively, where the convergence is through values of c such that $\mu + c \in \Omega$ and $e^{\beta c}/\sigma^2 \in \Lambda$. The model must be infinitely divisible if $\beta c \to -\infty$.

Proof. Consider the case where V is regular at infinity. For fixed values of c, σ^2 and β, the left-hand side of (4.77) is a natural

exponential family with unit variance function
$$V_c(\mu) = \sigma^2 e^{-\beta c} V(\mu + c) \to c_0 \sigma^2 e^{\beta \mu} \quad \text{as} \quad c \to \infty.$$
We now show that the convergence is uniform in μ on compact subsets of $\mathbf{R}$. To this end, let $|\mu| \leq M$ and let μ_0 be such that
$$\left|V(\mu)e^{-\beta\mu} - c_0\right| < \varepsilon$$
for all $\mu > \mu_0$. Then for any $c > \mu_0 + M$
$$\begin{aligned}
\left|V(\mu+c)e^{-\beta c} - c_0 e^{\beta\mu}\right| &= e^{\beta\mu}\left|V(\mu+c)e^{-\beta(c+\mu)} - c_0\right| \\
&\leq M_0 \left|V(\mu+c)e^{-\beta(c+\mu)} - c_0\right| \\
&\leq M_0 \varepsilon,
\end{aligned}$$
where $M_0 = e^{\beta M} + e^{-\beta M}$. This shows that the convergence is uniform. The result now follows from the Mora convergence theorem. □

Extending the terminology from above, we see that the convergence in Theorem 4.10 may be either of central limit type or infinitely divisible type, depending on whether $c\beta$ goes to plus or minus infinity, respectively. In the case $\beta = 0$ the variance function is also of order zero. This case corresponds to the following corollary.

Corollary 4.11 *If the unit variance function V for the reproductive family $ED(\mu, \sigma^2)$ is asymptotically constant c_0 at infinity, then*
$$-c + \mathrm{ED}\left(\mu + c, \sigma^2\right) \xrightarrow{d} N(\mu, \sigma^2 c_0) \quad \text{as} \quad c \to \infty.$$

Example 4.8 Taking $\gamma = -1$ in the generalized inverse Gaussian distribution (Section 4.4.4), we obtain the asymptotic behaviour
$$V_{-1,\chi}(\mu) \sim \chi^2 \exp\left(\frac{2\mu}{\chi}\right) \quad \text{as} \quad \mu \to \infty.$$
The model is hence of order ∞ at infinity with coefficient $2/\chi$. By Theorem 4.10 we hence obtain
$$-c + \mathrm{ED}_{-1,\chi}\left(\mu + c, \sigma^2 e^{-\frac{2c}{\chi}}\right) \xrightarrow{d} \mathrm{Tw}_\infty\left(\mu, \chi^2 \sigma^2, \frac{2}{\chi}\right) \quad \text{as} \quad c \to \infty.$$
The convergence is of central limit type. □

4.6 Tweedie-Poisson mixtures

While the Tweedie class consists of mainly continuous models, we now introduce a class of Tweedie-Poisson mixtures that are suitable for positive count data.

4.6.1 Poisson mixtures

Consider a Poisson mixture defined as follows. Let $X \sim \mathrm{ED}^*(\varphi, \lambda)$ denote an additive exponential dispersion model with density
$$p_X(x; \varphi, \lambda) = c^*(x; \lambda) \exp\{\varphi x - \lambda \kappa(\varphi)\},$$
support $S \subseteq \mathbf{R}_0$, canonical parameter domain Φ with $\sup \Phi = \varphi_0 \leq 0$, and index set Λ. Define the conditional distribution of Z given X by
$$Z|X = x \sim \mathrm{Po}\left(xe^\theta\right),$$
where $\theta \in \mathbf{R}$ is a parameter. In the case where $x = 0$ we define Po(0) as the degenerate distribution at zero.

The joint density of X and Z is then
$$\begin{aligned}p_{X,Z}(x, z; \theta, \varphi, \lambda) &= \frac{x^z e^{\theta y}}{z!} c^*(x; \lambda) \exp\{-xe^\theta + \varphi x - \lambda \kappa(\varphi)\} \\ &= \frac{e^{\theta y}}{z!} x^z c^*(x; \lambda) \exp\{\psi x - \lambda \kappa(\psi + e^\theta)\}\end{aligned}$$
where $\psi = \varphi - e^\theta$. Hence the marginal probability function of Z is, for $z = 0, 1, \ldots$,
$$\begin{aligned}p_Z(z; \theta, \varphi, \lambda) &= \frac{e^{\theta y}}{z!} \int x^z c^*(x; \lambda) \exp\{\psi x - \lambda \kappa(\psi + e^\theta)\} \, dx \\ &= \frac{m(z; \psi, \lambda)}{z!} \exp\{\theta y - \lambda \kappa(\psi + e^\theta)\}. \quad (4.78)\end{aligned}$$
Here $m(z; \psi, \lambda)e^{-\lambda \kappa(\theta)}$ is the zth moment of X, or
$$\begin{aligned}m(z; \psi, \lambda) &= \int x^z c^*(x; \lambda) e^{\psi x} \, dx \\ &= \frac{\partial^z}{\partial \psi^z} \int c^*(x; \lambda) e^{\psi x} \, dx \\ &= \frac{\partial^z}{\partial \psi^z} e^{\lambda \kappa(\psi)}.\end{aligned}$$
Then (4.78) is a new additive exponential dispersion model for each fixed value of $\psi \leq \varphi_0$ with canonical parameter θ, canonical parameter domain $\log(\Phi - \psi)$ and unit cumulant function $\theta \mapsto \kappa(\psi + e^\theta)$. It has mean domain $\mathbf{R}_+$.

Let us denote the unit variance function of $\mathrm{ED}^*(\varphi, \lambda)$ by V, the mean domain by Ω and the mean value mapping by τ. Then the

mean of Z is
$$\mu = \tau\left(\psi + e^\theta\right)e^\theta.$$
Let $h_\psi(\mu)$ denote the inverse of the function $z \mapsto z\tau(\psi + z)$. The unit variance function V_ψ of the mixture (4.78) may then be expressed as follows, for $\mu > 0$:
$$\begin{aligned} V_\psi(\mu) &= \tau\{\psi + h_\psi(\mu)\}h_\psi(\mu) + \tau'\{\psi + h_\psi(\mu)\}h_\psi^2(\mu) \\ &= \mu + \tau'\left[\tau^{-1}\left\{\frac{\mu}{h_\psi(\mu)}\right\}\right]h_\psi^2(\mu) \\ &= \mu + V\left\{\frac{\mu}{h_\psi(\mu)}\right\}h_\psi^2(\mu). \end{aligned}$$

The second term of the last expression is positive, so that the mixture distribution is overdispersed compared with the Poisson distribution.

4.6.2 Tweedie case

General

In the Tweedie case we assume that
$$X \sim \mathrm{Tw}_p\left\{\lambda\tau_p(\varphi), \lambda^{1-p}\right\} = \mathrm{Tw}_p^*(\varphi, \lambda)$$
for some $p > 1$. The case $p = 1$ will be considered separately below. Because of the scale transformation property of the Tweedie model, the parameters θ and ψ are not identifiable in this case. It is convenient to choose $\psi = -1$, which implies that θ is nonpositive, and strictly negative if $1 < p < 2$, because now $\varphi_0 = 0$. The marginal distribution for Z, given by the appropriate form of (4.78), is an additive exponential dispersion model, which we denote by $Z \sim \mathrm{TP}_p^*(\theta, \lambda)$. We call it the **Tweedie-Poisson mixture distribution*** (Hougaard, Lee and Whitmore, 1996). It may be considered a discrete-data analogue of the Tweedie class. We now study some basic properties of this class of distributions.

We first consider the unit variance function. Let $z = h_p(\mu) \in (0,1)$ denote the solution to the equation $\mu = z\tau_p(z-1)$ for $\mu > 0$. The mean is then
$$\mu = \tau_p\{h_p(\mu) - 1\}h_p(\mu).$$

---
* Poisson mixtures are sometimes called *compound Poisson distributions*, but we use this terminology for a different purpose here, see Section 4.2.4.

Using $V(\mu) = \mu^p$ and denoting the unit variance function of (4.78) by V_p (not to be confused with (4.1)), we obtain, for $\mu > 0$,

$$V_p(\mu) = \mu + \mu^p h_p^{2-p}(\mu).$$

The variance of Z is then

$$\operatorname{var} Z = \lambda V_p\left(\frac{\mu}{\lambda}\right) = \mu + \lambda^{1-p}\mu^p h_p^{2-p}\left(\frac{\mu}{\lambda}\right).$$

The convergence properties of this unit variance function are given in the following proposition.

Proposition 4.12 *Let $p > 1$. The unit variance function V_p is then regular of order 1 at zero and regular of order p at infinity.*

Proof. The function $h_p(\mu)$ is strictly increasing and maps $(0, \infty)$ onto $(0, 1)$. This implies that V_p is of order p at infinity. It is easy to show that h_p satisfies the asymptotic relation

$$h_p(\mu) \sim \frac{\mu}{\tau_p(-1)} \quad \text{as} \quad \mu \to 0.$$

This implies that V_p is of order 1 at zero. □

The proposition shows that the Tweedie-Poisson unit variance function interpolates between Poisson behaviour at zero and Tweedie behaviour at infinity. Figure 4.9 shows some plots of the unit variance function for the Tweedie-Poisson mixture.

The unit deviance may be expressed as follows:

$$d(y; \mu) = 2\left\{y \log \frac{h_p(y)}{h_p(\mu)} - \psi_p(y) + \psi_p(\mu)\right\},$$

where $\psi_p(\mu) = \kappa_p\{h_p(\mu) - 1\}$. Hence, the unit deviance and the unit variance function both involve $h_p(\mu)$ in their expressions.

The Tweedie-Poisson mixtures include a few cases of known distributions. First of all, $\mathrm{TP}_2^*(\theta, \lambda)$ is the negative binomial distribution. There are two further cases where h_p has closed-form expressions. We now consider these along with the limiting case $p = 1$.

Inverse Gaussian-Poisson mixture

For the distribution $\mathrm{TP}_3^*(\theta, \lambda)$, where the mixing distribution is the inverse Gaussian distribution, we obtain the following expression for h_3 for $\mu > 0$:

$$h_3(\mu) = \mu\left(\sqrt{2 + \mu^2} - \mu\right).$$

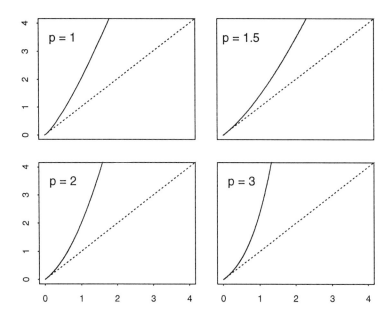

Figure 4.9. *Some Tweedie-Poisson mixture unit variance functions.*

This leads to the following unit variance function, for $\mu > 0$:

$$V_3(\mu) = \mu + \frac{\mu^3}{2} + \frac{\mu^2}{2}\sqrt{2+\mu^2}.$$

This unit variance function is of Letac form. The distribution $\text{TP}_3^*(\theta, \lambda)$ has been considered by Holla (1966), Sankharan (1968) and Sichel (1971).

Compound Poisson-Poisson mixture

In the special case $1 < p < 2$ the mixing distribution is a compound Poisson distribution, and the corresponding Tweedie-Poisson mixtures have the special feature of a surplus of zeros because of the positive probability at zero of the compound Poisson distribution. These distributions may hence be useful for data that exhibit a surplus of zeros, a feature that is seen, for example, in certain bacterial count data.

For $p = 1.5$, we obtain the following expression for $h_{1.5}$ for $\mu > 0$:

$$h_{1.5}(\mu) = \frac{\mu + 2 - 2\sqrt{1+\mu}}{\mu}.$$

The unit variance function then takes the form

$$V_{1.5}(\mu) = \mu + \mu\sqrt{\mu + 2 - 2\sqrt{1+\mu}}$$

for $\mu > 0$. We note that this unit variance function is of order 1.5 at infinity, corresponding to an asymptotic behaviour that is an intermediate between the Poisson and the negative binomial.

Neyman type A distribution

The limiting case $p = 1$, which we denote $\text{TP}_1^*(\theta, \lambda)$, corresponds to a Poisson mixing distribution. This distribution is known as the Neyman type A distribution (see Johnson, Kotz and Kemp, 1992). In this case the canonical parameter domain is $\mathbf{R}$, and the equation $\mu = z\tau_1(z-1) = z\exp(z-1)$ has a solution $z = h_1(\mu) > 0$. The unit variance function is

$$V_1(\mu) = \mu + \mu h_1(\mu)$$

for $\mu > 0$. This unit variance function is regular of order 1 at zero, but is not regular of any order at infinity. The latter fact is remarkable, for being the only such case among the examples we have studied so far.

4.7 Notes

The Tweedie models are named after M.C.K. Tweedie, who introduced this class of models in the paper Tweedie (1984). However, a number of authors have studied this idea independently of Tweedie, see for example Morris (1981), Hougaard (1986) and Bar-Lev and Enis (1986). Jørgensen (1987a) seems to have been the first to describe the full domain for the parameter p, including the non-steep case $p < 0$. The Tweedie class generalizes the so-called non-central chi-square distribution with zero degrees of freedom, see Jones (1987), Siegel (1985) and references therein, and also Exercise 4.11. The Tweedie additive process was called a *Hougaard process* by Lee and Whitmore (1993). Exponential unit variance functions were mentioned by McCullagh (1983), and the connection with extreme stable distributions was mentioned in the reply to the discussion of Jørgensen (1987a). The characterization of the Tweedie class in terms of scale transformations and translations is due to Jørgensen (1992). The Tweedie convergence results are due to Jørgensen and Martínez (1997) and Jørgensen, Martínez and Tsao (1994), see also Daniels (1954), Jensen (1988, 1989, 1992)

EXERCISES

and Barndorff-Nielsen and Klüppelberg (1992). Hougaard (1995) considered the saddlepoint approximation to the positive stable distributions.

The inverse Gaussian distribution is also associated with Tweedie's name, due to his pioneering study of its statistical properties (Tweedie, 1957). Seshadri (1994) provides an extensive study of the inverse Gaussian distribution from the point of view of exponential families.

4.8 Exercises

Exercise 4.1 Complete the proof of Theorem 4.1 via the following steps.

1. Show that the relation $V(c\mu) = g(c)V(\mu) \quad \forall c > 0, \mu > 0$ may be written in the form
$$V(c\mu) = V(c)V(\mu).$$

2. Re-express this relation in the form
$$f(x+y) = f(x) + f(y),$$
where $f(x) = \log V(e^x)$.

3. Show that the derivative of f is constant, and equal to p, say, and conclude that $V(\mu) = \mu^p$.

Exercise 4.2 Verify the scale transformation property
$$c\mathrm{Tw}_1(\mu, \sigma^2) = \mathrm{Tw}_1(c\mu, c\sigma^2)$$
in the case of the Poisson distribution in its reproductive form, $\mathrm{Tw}_1(\mu, \sigma^2) = \sigma^2 \mathrm{Po}(\mu/\sigma^2)$.

Exercise 4.3 Let $\mathrm{ED}(\mu, \sigma^2)$ be an exponential dispersion model satisfying for all $c > 0$
$$c\mathrm{ED}(\mu, \sigma^2) = \mathrm{ED}(c\mu, \sigma^2).$$
Show that if Ω contains a positive value, then $\mathrm{ED}(\mu, \sigma^2)$ is the gamma distribution.

Exercise 4.4 Let $\mathrm{ED}(\mu, \sigma^2)$ be an exponential dispersion model satisfying the following relation for all $c \in \mathbf{R}$:
$$c + \mathrm{ED}(\mu, \sigma^2) = \mathrm{ED}(c + \mu, \sigma^2).$$
Show that $\mathrm{ED}(\mu, \sigma^2)$ is the normal distribution.

Exercise 4.5 Show that the inverse of the relation (4.8) is

$$\text{Tw}_p(\mu, \sigma^2) = \text{Tw}_p^* \left[\tau_p^{-1} \left\{ \mu \sigma^{\frac{2}{p-1}} \right\}, \sigma^{\frac{2}{1-p}} \right].$$

Exercise 4.6 (Inverse Gaussian distribution) Let the function $F(\,\cdot\,; \xi, \lambda)$ be defined by

$$F(y; \xi, \lambda) = \Phi\left(\xi\sqrt{\lambda y} - \sqrt{\lambda/y}\right) + e^{2\lambda \xi} \Phi\left(-\xi\sqrt{\lambda y} - \sqrt{\lambda/y}\right)$$

for $y > 0$ and let F be zero for $y \leq 0$, where Φ denotes the standard normal distribution function. Show that F is a distribution function for $\xi \geq 0$, $\lambda > 0$, using the following steps.

1. Show that $F(\infty; \xi, \lambda) = 1$.
2. Show that the right limit of $F(\,\cdot\,; \xi, \lambda)$ at zero is zero.
3. Show that the derivative of F with respect to y is

$$F'(y; \xi, \lambda) = \sqrt{\frac{\lambda}{2\pi y^3}} \exp\left\{-\frac{\lambda}{2}(\xi^2 y + y^{-1} - 2\xi)\right\},$$

and show that F is increasing.

4. Show that $F(\,\cdot\,; \xi, \lambda)$ defines a continuous distribution with support $\mathbf{R}_+$ and that $F'(y; \xi, \lambda)$ is a density.

Exercise 4.7 Show that if $Y \sim \text{IG}(\infty, \lambda^{-1})$, then $\lambda/Y \sim \chi^2(1)$.

Exercise 4.8 Find the exponential dispersion model corresponding to the unit variance function $V(\mu) = (\mu - \alpha)^3$, $\mu > \alpha$, where $\alpha \in \mathbf{R}$ is a given constant.

Exercise 4.9 Consider a variable $Y \sim \text{IG}(\mu, \sigma^2)$ following the inverse Gaussian distribution.

1. Calculate the five standard residuals for this distribution, and the modified deviance residual.
2. Determine whether the conditions (3.54) and (3.55) hold in this example.

Exercise 4.10 (Continuation of Exercise 4.9.) Investigate the r^* and Lugannani–Rice approximations to the inverse Gaussian distribution function, responding to the following questions.

1. Show, using the scale transformation property of the inverse Gaussian distribution, that the accuracy of the approximations depends only on the value of the parameter $\delta = \sigma^2 \mu$. Hint: Consider the distribution of Y/μ.

EXERCISES

2. Make a numerical investigation of the approximations, for a range of values of δ and y. Determine an empirical rule that says how small δ must be in order to obtain an accuracy of four significant digits. Hint: The inverse Gaussian distribution function was calculated in Exercise 4.6.

Exercise 4.11 (Noncentral chi-square) The noncentral chi-square distribution with ν degrees of freedom and noncentrality parameter λ has probability density function for $z > 0$ given by

$$p(z; \nu, \lambda) = \sum_{n=0}^{\infty} \frac{\left(\frac{z}{2}\right)^{\frac{\nu}{2}+n} \lambda^n}{x\Gamma\left(\frac{\nu}{2}+n\right) n!} e^{-\frac{z}{2}-\lambda}.$$

1. Show that this is a Poisson mixture of chi-square distributions, as follows:

$$Z|N = n \sim \chi^2(\nu + 2n)$$
$$N \sim \text{Po}(\lambda).$$

2. Generalize the noncentral chi-square distribution, by defining

$$Z|N = n \sim \text{Ga}^*(\theta, \lambda r + \rho n)$$
$$N \sim \text{Po}\left\{\lambda(-\theta)^{-\rho}\right\}.$$

Find the probability density function of Z, and show that it is an additive exponential dispersion model for $r, \rho > 0$ known. Show that both the noncentral chi-square and Tweedie compound Poisson models are special cases of this distribution.

3. Find the unit variance function of the model in the special case $\rho = 1$. Are there any other values of ρ for which the unit variance function has a closed-form expression?

4. For each value of r and ρ, determine if the variance function is regular at zero or infinity. In the affirmative cases, find the order.

Exercise 4.12 Show that the unit variance function of the generalized Poisson distribution (4.58) is $\mu(1+\mu)^2$ for $\mu > 0$. Hint: Differentiate the equation $\log \kappa(\theta) - \kappa(\theta) = \theta$ twice, and express the second derivative in terms of $\mu = \kappa'(\theta)$.

Exercise 4.13 Let $\kappa(\theta) = \theta \log \theta$, $\theta \geq 0$. Find the corresponding unit variance function, and find the corresponding exponential dispersion model.

Exercise 4.14 Find the saddlepoint approximation for the exponential dispersion model with unit variance function $V(\mu) = e^\mu$.

Exercise 4.15 Consider a strictly cubic unit variance function defined on $\Omega = \mathbf{R}_+$ by
$$V(\mu) = a\mu^3 + b\mu^2 + c\mu.$$
Show that this unit variance function is regular at zero and at infinity, and determine the two respective orders for all the combinations of a, b and c.

Exercise 4.16 Prove the form of the variance function (4.60).

Exercise 4.17 Let $Y \sim \mathrm{IG}(\mu, \sigma^2)$. Show, by direct calculation, that the density function of the variable $W = (Y - \mu)/\sigma$ converges to the density of the normal distribution $N(0, \mu^3)$ as $\sigma^2 \to 0$.

Exercise 4.18 Consider the variance function $V(\mu) = \left(1 + \mu^2\right)^{3/2}$ from Exercise 3.2. Show that V is regular of order 3 at infinity, and write down the corresponding convergence result.

Exercise 4.19 Simulate the inverse Gaussian process, using the method of Michael, Schucany and Hass (1976).

Exercise 4.20 Let $X \sim \mathrm{ED}^*(\varphi, \lambda)$ denote an additive exponential dispersion model with density
$$p_X(x; \varphi, \lambda) = c^*(x; \lambda) \exp\left\{\varphi x - \lambda \kappa(\varphi)\right\},$$
support $S \subseteq \mathbf{R}_0$, canonical parameter domain Φ and index set Λ. Define a normal mixture as follows. Let the conditional distribution of Y given X be
$$Y|X = x \sim N(\theta x, x),$$
where $\theta \in \mathbf{R}$ is a parameter. In the case where $x = 0$ we define $N(0, 0)$ as the degenerate distribution at zero. Examine the marginal distribution of Y along the same lines as the Tweedie-Poisson mixture of Section 4.6. In particular, examine the special case where X follows a Tweedie model.

Exercise 4.21 Classify the convergence results for the six Morris distributions in Chapter 3 into the three types 'central limit', 'infinitely divisible' and 'Tauber' and determine the corresponding orders. By $\mathrm{ED}(\mu, \sigma^2)$ being of 'order zero at a point m' we mean that $\mathrm{ED}(\mu, \sigma^2) - m$ is of order zero at zero. Prove that any of the models is regular of order zero at any point in the mean domain.

CHAPTER 5

Proper dispersion models

Proper dispersion models were introduced in Chapter 1. We now study their properties in more detail and introduce some further examples.

5.1 General dispersion models

5.1.1 Definitions

Let us recall the definition of unit deviances and proper dispersion models from Chapter 1. If $\Omega \subseteq C$ are two intervals with Ω open, a unit deviance $d(y; \mu)$ on $C \times \Omega$ is a non-negative function satisfying $d(y; y) = 0$. It is called regular if it is twice continuously differentiable and $\partial^2 d/\partial \mu^2(y; y) > 0$ for $y \in \Omega$. The unit variance function is defined on Ω by

$$V(\mu) = \frac{2}{\frac{\partial^2 d}{\partial \mu^2}(\mu; \mu)}. \tag{5.1}$$

It is convenient in the present chapter to work with the index parameter λ instead of the dispersion parameter $\sigma^2 = 1/\lambda$. In this notation, a **standard dispersion model** $\text{DM}(\mu, \sigma^2)$ is defined by

$$p(y; \mu, \lambda) = c(y; \lambda) \exp\left\{-\frac{\lambda}{2} d(y; \mu)\right\}, \quad y \in S, \tag{5.2}$$

where d is a unit deviance (not necessarily regular). Here S denotes the set of realizable values for the family, C is the convex support and the density is with respect to Lebesgue measure. We let Λ denote the domain for λ, called the **index set**. Throughout the chapter it is understood that 'dispersion model' refers to the continuous reproductive case.

If d is regular with unit variance function V, and $\Omega = S = C$, a **regular proper dispersion model** $Y \sim \text{PD}(\mu, \sigma^2)$ is defined by

$$p(y; \mu, \lambda) = c(\lambda) V^{-\frac{1}{2}}(y) \exp\left\{-\frac{\lambda}{2} d(y; \mu)\right\}, \quad y \in \Omega. \quad (5.3)$$

In a slight abuse of notation, we let the $c(y; \lambda)$ of the density (5.2) factorize as $c(y; \lambda) = c(\lambda) V^{-1/2}(y)$ in the case of the regular proper dispersion model density (5.3). For a regular proper dispersion model, the index set Λ is an interval that is unbounded to the right.

From now on we shall not restrict λ to positive values, although so far the only known example where negative values of λ occur is Leipnik's distribution, introduced in Section 5.3.2 below. It is useful to continue to let $\mathrm{DM}(\mu, \sigma^2)$ and $\mathrm{PD}(\mu, \sigma^2)$ refer to (5.2) and (5.3), respectively, with $\lambda = 1/\sigma^2$ restricted to positive values.

In the definition of regular proper dispersion models, it may seem overly restrictive to assume that d is a unit deviance and to assume that the factor $V^{-1/2}(y)$ in (5.3) is defined in terms of d, instead of being an arbitrary function of y. However, we now show that these restrictions are, to a large extent, only apparent. We hence introduce the following definition, based on which we shall be able to substantiate this claim.

Definition 5.1 A **general dispersion model** *with parameters* $\theta \in \Theta$ *and* $\lambda \in \Lambda$ *is a family of distributions for* Y *with probability density functions of the form*

$$p(y; \theta, \lambda) = c(y; \lambda) \exp\{\lambda t(y; \theta)\}, \quad y \in S, \quad (5.4)$$

where $c \geq 0$ *and* t *are suitable functions,* S *denotes the set of realizable values of the family,* Θ *is an interval and* Λ *is unbounded to the right. If* c *factorizes as follows:*

$$c(y; \lambda) = c(\lambda) b(y) \quad \forall y \in S, \lambda \in \Lambda \quad (5.5)$$

for suitable functions c *and* b, *we call (5.4) a* **general proper dispersion model**. *If* t *takes the form*

$$t(y; \theta) = \mathbf{s}^\top(y) \boldsymbol{\alpha}(\theta) \quad (5.6)$$

for appropriate vector functions $\mathbf{s}(y)$ *and* $\boldsymbol{\alpha}(\theta)$ *in* $\mathbf{R}^k$, *we call (5.4) an* **exponential family dispersion model**.

Note that an exponential dispersion model is an example of an exponential family dispersion model, corresponding to the form $\mathbf{s}(y) = (y, 1)^T$. However, an exponential family dispersion model need not be an exponential dispersion model, as illustrated by several of the examples in Section 5.3.

Pseudo-dispersion models

Let us recall another definition from Chapter 1. Let $d(y;\mu)$ be a given regular unit deviance on $\Omega \times \Omega$ with unit variance function V. Suppose that the integral

$$\frac{1}{c_0(\mu,\lambda)} = \int_\Omega V^{-\frac{1}{2}}(y)\exp\left\{-\frac{\lambda}{2}d(y;\mu)\right\} dy \qquad (5.7)$$

is finite for all $\mu \in \Omega$ and λ in some set Λ that is unbounded to the right. The corresponding **renormalized saddlepoint approximation** is defined as the family of distributions with probability density functions on Ω given by

$$p(y;\mu,\lambda) = c_0(\mu,\lambda) V^{-\frac{1}{2}}(y)\exp\left\{-\frac{\lambda}{2}d(y;\mu)\right\}. \qquad (5.8)$$

We refer to $c_0(\mu,\lambda)$ as a **normalizing constant**; when it appears in a formula like (5.8), it is understood that it is defined by (5.7), in order to make (5.8) a probability density function.

We now generalize renormalized saddlepoint approximations, inspired by the notion of a saddlepoint approximation being uniform on compacts.

Definition 5.2 *Let $d(y;\mu)$ be a unit deviance on $C \times \Omega$ with unit variance function V. A family of distributions with probability density functions on the set of realizable values S given by*

$$p(y;\mu,\lambda) = c(y;\mu,\lambda)\exp\left\{-\frac{\lambda}{2}d(y;\mu)\right\} \qquad (5.9)$$

is called a **pseudo-dispersion model** *if c satisfies*

$$\lambda^{-\frac{1}{2}}c(y;\mu,\lambda) \to \{2\pi V(y)\}^{-\frac{1}{2}} \quad \text{as} \quad \lambda \to \infty,$$

for all $y \in C$, and if, for each μ, the convergence is uniform in y on compact subsets of Ω.

By Theorem 1.3, a renormalized saddlepoint approximation is also a pseudo-dispersion model. The class of pseudo-dispersion models extends the definition of reproductive dispersion models by allowing $c(y;\mu,\lambda)$ to depend on μ.

5.1.2 Properties

General dispersion models

We now consider some properties of general dispersion models. Note that c and t are not uniquely determined from (5.4), because, for an arbitrary function u, the following:

$$c(y;\lambda)e^{\lambda u(y)} \exp\left[\lambda\{t(y;\theta) - u(y)\}\right] \qquad (5.10)$$

represents the same density, but with t replaced by $t(y;\theta) - u(y)$ and c by $c(y;\lambda)\exp\{\lambda u(y)\}$. In particular, it is conceivable that the factorization (5.5), when it holds, may hold only for certain choices of t.

In some cases, it is possible to choose u in such a way that $2\{u(y) - t(y;\theta)\}$ is a unit deviance. The following definition, which alludes to the concept of a *yoke* in Exercise 1.2, formalizes this idea.

Definition 5.3 *The function t is called* **yokable** *if it satisfies:*

1. $\sup_{\theta\in\Theta} t(y;\theta) < \infty$ *for all $y \in C$.*
2. *There exists an open interval $\Omega \subseteq C$ such that, for all $y \in \Omega$, the mode point $\hat{\theta}(y)$ of $t(y;\cdot)$ is unique and $\hat{\theta}(\cdot)$ is a one-to-one mapping of Ω onto $\operatorname{int}\Theta$.*

If t is yokable, we define the function $d: C \times \Omega \to \mathbf{R}$ by

$$d(y;\mu) = 2\left[\hat{t}(y) - t\left\{y;\hat{\theta}(\mu)\right\}\right], \qquad (5.11)$$

where $\hat{t}(y)$ is defined for $y \in C$ by

$$\hat{t}(y) = \sup_{\theta\in\Theta} t(y;\theta);$$

in particular $\hat{t}(y) = t\left\{y;\hat{\theta}(y)\right\}$ on Ω. We then define the parameter $\mu \in \Omega$ as the solution of $\hat{\theta}(\mu) = \theta$, which is a reparametrization of θ on $\operatorname{int}\Theta$. Furthermore, d is a unit deviance on $C \times \Omega$. As we did for exponential dispersion models, we assume that μ is defined by continuity at the end points of Θ, taking infinite values if necessary.

For an exponential dispersion model of the form

$$p(y;\theta,\lambda) = c(y;\lambda)\exp\left[\lambda\{\theta y - \kappa(\theta)\}\right],$$

we have $t(y;\theta) = y\theta - \kappa(\theta)$. This t is yokable, and the corresponding unit deviance (5.11) is the usual one for exponential dispersion models.

In the general case, expressing the density in terms of the unit deviance (5.11), we obtain a standard dispersion model with density

$$p(y;\theta,\lambda) = c(y;\lambda)\exp\left\{\lambda\hat{t}(y) - \frac{\lambda}{2}d(y;\mu)\right\}, \qquad (5.12)$$

where the new c function is

$$c(y;\lambda)\exp\left\{\lambda\hat{t}(y)\right\}. \qquad (5.13)$$

GENERAL DISPERSION MODELS

Note that this construction is unique, in the sense that it would give the same result if applied to (5.10), independently of the choice of the function u. Whenever t is yokable, there is hence no loss of generality by assuming, in (5.4), that $-2t$ is a unit deviance, i.e. that the model (5.4) is a standard dispersion model.

How restrictive is it to assume that t is yokable? Item 1. of Definition 5.3 is satisfied for most reasonable models, because $\lambda t(y; \cdot)$ is a log likelihood when λ is known. Item 2. implies that θ is a reparametrization of a position parameter (in the terminology of Chapter 1). Although this is a fairly strong condition, we know of no examples of general dispersion models that are not standard. It is hence an open question whether or not t may always be assumed to be yokable.

Proper dispersion models

Consider a general proper dispersion model with density

$$p(y;\theta,\lambda) = c(\lambda)b(y)\exp\{\lambda t(y;\theta)\}, \quad y \in S, \tag{5.14}$$

where $\theta \in \Theta$, $\lambda \in \Lambda$. A crucial property of this model is that the integral

$$\frac{1}{c(\lambda)} = \int_S b(y)\exp\{\lambda t(y;\theta)\}\,dy \tag{5.15}$$

does not depend on θ. A closely related fact is that for θ known the model (5.14) is an exponential family of order 1 with canonical statistic $T = t(Y;\theta)$ and canonical parameter λ. Note the contrast with exponential family dispersion models, which are exponential families for λ known. In the following, a **pivot** denotes a function of Y and θ whose distribution under θ does not depend on the value of θ.

Proposition 5.1 *The function $T = t(Y;\theta)$ is a pivot when λ is known.*

Proof. The moment generating function of T is, for $\lambda + s \in \Lambda$,

$$\begin{aligned} M_T(s;\theta,\lambda) &= \int_S e^{st(y;\theta)}c(\lambda)b(y)e^{\lambda t(y;\theta)}\,dy \tag{5.16} \\ &= c(\lambda)\int_S b(y)\exp\{(s+\lambda)t(y;\theta)\}\,dy \\ &= \frac{c(\lambda)}{c(\lambda+s)}. \end{aligned}$$

Since the domain of a moment generating function is an interval, we find that Λ is an interval, and, being unbounded to the right,

Λ is not degenerate. Hence, the moment generating function of T characterizes the distribution of T. But since M_T depends on (θ, λ) via λ only, we conclude that T is a pivot for λ known. $\square$

The property that T is a pivot is crucial in the following. For example, it may be useful for simulating values from a proper dispersion model, by first simulating a value of T and then inverting t to obtain values of y, see Michael, Schucany and Hass (1976) and McCullagh (1989) for details.

Now if T is a pivot, then so is every function only of T. The following theorem, due to S.L. Lauritzen, shows a slightly more general version of this result.

Lemma 5.2 *Consider the general proper dispersion model (5.14). Then for any measurable function f the integral*

$$\int_S b(y) \exp\left[\lambda f\{t(y; \theta)\}\right] dy$$

is independent of θ.

Proof. The proof relies on (5.15), which says that the Laplace transform of $t(\,\cdot\,; \theta)$ with respect to the measure $b(y)\,dy$ does not depend on θ. But then the same holds for any function of $t(y; \theta)$, proving the lemma. $\square$

The lemma shows that any given general proper dispersion gives rise to a large class of related general proper dispersion models given by

$$p(y; \theta, \lambda) = c_f(\lambda) b(y) \exp\left[\lambda f\{t(y; \theta)\}\right],$$

because, by the lemma, the integral

$$\frac{1}{c_f(\lambda)} = \int_S b(y) \exp\left[\lambda f\{t(y; \theta)\}\right] dy$$

does not depend on θ.

Consider now the special case of a regular proper dispersion model given by

$$p(y; \mu, \lambda) = c(\lambda) V^{-\frac{1}{2}}(y) \exp\left\{-\frac{\lambda}{2} d(y; \mu)\right\}.$$

In this case it is convenient to think of f as a transformation of the unit deviance d. Thus, suppose that f is twice differentiable, maps $\mathbf{R}_+$ into $\mathbf{R}_+$, and satisfies $f(0) = 0$ and $f'(0) > 0$. Then $f\{d(y; \mu)\}$ is a regular unit deviance with unit variance function $V(\mu)/f'(0)$.

Hence, by Lemma 5.2, the following:

$$p(y;\mu,\lambda) = c_f(\lambda) V^{-\frac{1}{2}}(y) \exp\left[-\frac{\lambda}{2} f\{d(y;\mu)\}\right],$$

is a new regular proper dispersion model, where c_f is a normalizing constant. In some cases we may give a more explicit interpretation of the transformed model, as we shall see in Section 5.4.

Pseudo-dispersion models

The main property of a pseudo-dispersion model is that it has a saddlepoint approximation,

$$p(y;\mu,\lambda) \sim \sqrt{\frac{\lambda}{2\pi}} V^{-\frac{1}{2}}(y) \exp\left\{-\frac{\lambda}{2} d(y;\mu)\right\}.$$

This implies, in turn, that the model is asymptotically normal, as shown in the next theorem.

Theorem 5.3 *Let Y follow the pseudo-dispersion model (5.9) with parameters μ and λ. Then*

$$\sqrt{\lambda}(Y-\mu) \xrightarrow{d} N\{0, V(\mu)\} \quad \text{as} \quad \lambda \to \infty.$$

The proof is similar to the proof of Theorem 1.5, and is left to the reader. Note that a more general convergence result as in Theorem 1.5, where μ has the form $\mu_0 + \mu/\sqrt{\lambda}$, is not available for pseudo-dispersion models without further assumptions on c, see also Exercise 1.12.

5.1.3 Barndorff-Nielsen's formula

The key to understanding the structure of general proper dispersion models lies in Barndorff-Nielsen's (1983, 1988) p^* formula, which we now consider. Suppose that we have a general proper dispersion model

$$p(y;\theta,\lambda) = c(\lambda) b(y) \exp\{\lambda t(y;\theta)\}, \quad y \in S,$$

such that t is yokable. Following the above notation, we may write the density in the form

$$p(y;\theta,\lambda) = c(\lambda) b(y) \exp\left\{\lambda \hat{t}(y) - \frac{\lambda}{2} d(y;\mu)\right\}. \tag{5.17}$$

Note that we have assumed a factorization of the original c, rather than of the c defined by (5.13). We furthermore assume that d is a regular unit deviance with unit variance function V, and that $\Omega = S = C$.

Under these assumptions, we now consider conditions for which (5.17) is a regular proper dispersion model. As equation (5.17) indicates, this question involves both of the two functions $b(y)$ and $\hat{t}(y)$.

Barndorff-Nielsen's formula provides an approximation to the conditional distribution of the maximum likelihood estimator for a given statistical model, given an ancillary statistic. Here we apply the formula in the case of a single observation from the general proper dispersion model (5.17) with λ known, in which case the ancillary statistic is degenerate, and the formula provides an approximation to the marginal distribution of the maximum likelihood estimator.

Since the maximum likelihood estimate of μ for λ fixed is $\hat{\mu}(y) = y$, Barndorff-Nielsen's formula hence provides an approximation to the marginal distribution of Y. This approximation is defined by

$$p\{y; \theta(\mu), \lambda\} \sim p_0(y; \mu, \lambda), \tag{5.18}$$

where p_0 is the renormalized saddlepoint approximation corresponding to the unit deviance d,

$$p_0(y; \mu, \lambda) = c_0(\mu, \lambda) V^{-\frac{1}{2}}(y) \exp\left\{-\frac{\lambda}{2} d(y; \mu)\right\}. \tag{5.19}$$

We say that Barndorff-Nielsen's formula is **exact** if the two sides of (5.18) are identical for all y and μ in Ω and all λ in Λ.

For standard dispersion models, we know that the saddlepoint approximation implies asymptotic normality of Y for λ large. We are interested in the cases where the same holds for general proper dispersion models. Barndorff-Nielsen's formula may be considered a refinement of the conventional normal approximation to the distribution of the maximum likelihood estimator. In the next theorem, we investigate the case where Barndorff-Nielsen's formula applies asymptotically for λ large.

Theorem 5.4 *Consider a general proper dispersion model such that t is yokable, the corresponding unit deviance d is regular with unit variance function V, and $\Omega = C = S$. Then the following three statements are equivalent:*

1. *Barndorff-Nielsen's formula is exact for all λ.*
2. *Barndorff-Nielsen's formula is asymptotically exact in the sense that the ratio p_0/p tends to 1 as $\lambda \to \infty$ for all y, μ in Ω.*

GENERAL DISPERSION MODELS

3. The function $\hat{t}(y)$ is constant on Ω, and

$$b(y) \propto V^{-\frac{1}{2}}(y). \tag{5.20}$$

When these statements hold, the normalizing constant $c_0(\mu, \lambda)$ does not depend on μ, and satisfies

$$c_0(\mu, \lambda) \propto c(\lambda) \exp\left\{\lambda \hat{t}(\mu)\right\}, \tag{5.21}$$

and furthermore

$$c(\lambda) \sim \sqrt{\frac{\lambda}{2\pi}} \exp\left\{-\lambda \hat{t}(\mu)\right\} \quad \text{as} \quad \lambda \to \infty. \tag{5.22}$$

Proof. Obviously 1. implies 2. We now show that 2. implies 3. Let us consider the ratio of the two sides of (5.18), namely

$$\frac{p_0(y; \mu, \lambda)}{p(y; \hat{\theta}(\mu), \lambda)} = \frac{c_0(\mu, \lambda) V^{-\frac{1}{2}}(y)}{c(\lambda) b(y) \exp\left\{\lambda \hat{t}(y)\right\}}. \tag{5.23}$$

Assume that 2. holds and let y_1 and y_2 be two values of y in Ω. Since (5.23) converges to 1, then so does the corresponding ratio of the two values of (5.23). This implies that, as $\lambda \to \infty$,

$$\frac{b(y_2) V^{-\frac{1}{2}}(y_1)}{b(y_1) V^{-\frac{1}{2}}(y_2)} \exp\left[\lambda \left\{\hat{t}(y_2) - \hat{t}(y_1)\right\}\right] \to 1. \tag{5.24}$$

Hence $\hat{t}(y)$ is constant on Ω, and since the ratio in (5.24) is then identically equal to 1, it follows that b is proportional to $V^{-1/2}$, proving (5.20).

That 3. implies 1. follows by noting that, if 3. holds, then (5.23) is constant. But since (5.23) is the ratio of two densities, the ratio must be 1, implying that Barndorff-Nielsen's formula is exact.

If the three statements hold, we obtain (5.21) from the fact that (5.23) is constant, noting that $\hat{t}(\mu) = \hat{t}(y)$. Since $\hat{t}(\mu)$ does not depend on μ, (5.21) implies that the same is the case for $c_0(\mu, \lambda)$. Finally, (5.22) follows from Theorem 1.3. □

Corollary 5.5 *Consider a general proper dispersion model satisfying the regularity conditions of Theorem 5.4 with b being continuous at $y = \mu$. If $\hat{t}(y)$ is constant, in particular if $-2t$ is a unit deviance, then Barndorff-Nielsen's formula is exact.*

Proof. Because of the constancy of $\hat{t}(y)$, we may write the integral (5.15) as follows:

$$\frac{1}{c(\lambda)} = \exp\left\{\lambda \hat{t}(\mu)\right\} \int_{\Omega} b(y) \exp\left\{-\frac{\lambda}{2} d(y; \mu)\right\} dy. \tag{5.25}$$

By the Laplace approximation to the integral (5.25), we obtain

$$c(\lambda) \sim \sqrt{\frac{\lambda}{2\pi}} \frac{K^{\frac{1}{2}}(\mu)}{b(\mu)} \exp\left\{-\lambda \hat{t}(\mu)\right\}, \qquad (5.26)$$

where

$$K(\mu) = \frac{\partial^2 d}{2\partial y^2}(\mu; \mu) = \frac{1}{V(\mu)}.$$

Here we have used the definition of V and Lemma 1.1. Now, (5.26) shows that $K^{1/2}(\mu)/b(\mu)$ is constant and hence $b(\mu) \propto V^{-1/2}(\mu)$. Consequently, item 3. of Theorem 5.4 is satisfied, and so Barndorff-Nielsen's formula is exact. □

The above result shows that Barndorff-Nielsen's formula is exact in a considerable range of cases, and the criterion of Corollary 5.5 is very easy to check. Under exactness, the density (5.17) is a regular proper dispersion model. We have found no examples where exactness does not hold when the regularity conditions of Theorem 5.4 are satisfied. There is hence little, if any, loss of generality by restricting attention to regular proper dispersion models, as we did in Chapter 1, and as we shall do in most of what follows. The main exceptions to this rule are the transformational dispersion models introduced below, where the unit deviance is not necessarily regular.

5.2 Construction of proper dispersion models

We shall now see that the renormalized saddlepoint approximation provides a useful method for constructing new proper dispersion models. This method works particularly well for transformational dispersion models, which generalize the location-dispersion models introduced in Chapter 1.

5.2.1 Renormalized saddlepoint approximations

The above analysis based on Barndorff-Nielsen's formula implies that the regular proper dispersion model given by

$$p(y; \mu, \lambda) = c(\lambda) V^{-\frac{1}{2}}(y) \exp\left\{-\frac{\lambda}{2} d(y; \mu)\right\}, \quad y \in \Omega, \qquad (5.27)$$

is essentially the most general form of general proper dispersion model possible under the assumption that d is a regular unit deviance with unit variance function V. This implies that the unit

CONSTRUCTION OF PROPER DISPERSION MODELS 185

deviance holds the key to the shape of the densities in the model, because the unit variance function is defined in terms of d, and c is in turn defined from d and V by

$$\frac{1}{c(\lambda)} = \int_\Omega V^{-\frac{1}{2}}(y) \exp\left\{-\frac{\lambda}{2}d(y;\mu)\right\} dy. \tag{5.28}$$

The model may hence be reconstructed from knowledge of the unit deviance only. This suggests a method for constructing new regular proper dispersion models, which we shall now consider.

Let $d(y;\mu)$ be a given regular unit deviance on $\Omega \times \Omega$ with unit variance function V. Suppose that the integral

$$\frac{1}{c_0(\mu,\lambda)} = \int_\Omega V^{-\frac{1}{2}}(y) \exp\left\{-\frac{\lambda}{2}d(y;\mu)\right\} dy \tag{5.29}$$

is finite for a set of λ-values that is unbounded to the right, and let

$$p(y;\mu,\lambda) = c_0(\mu,\lambda)V^{-\frac{1}{2}}(y) \exp\left\{-\frac{\lambda}{2}d(y;\mu)\right\}, \quad y \in \Omega, \tag{5.30}$$

be the corresponding renormalized saddlepoint approximation. We say that the unit deviance d is **proper** if $c_0(\mu,\lambda)$ depends on λ only and not on μ. If d is proper, then (5.30) takes the form

$$p(y;\mu,\lambda) = c(\lambda)V^{-\frac{1}{2}}(y) \exp\left\{-\frac{\lambda}{2}d(y;\mu)\right\}, \tag{5.31}$$

where $c(\lambda) = c_0(\mu,\lambda)$. We hence obtain a regular proper dispersion model.

It is relatively easy to construct new unit deviances. For example, if $t(y;\theta)$ is a one-parameter log likelihood which is yokable, then formula (5.11) provides a new unit deviance. If such a unit deviance is regular and proper, a new regular proper dispersion model may be constructed via (5.31). In particular, the class of regular proper dispersion models may be defined as the subset of the class of renormalized saddlepoint approximations corresponding to proper unit deviances. By finding new unit deviances and checking if they are proper, we hence have a method for constructing new regular proper dispersion models.

The crucial step in this construction is to check whether or not the integral (5.29) depends on μ. By differentiating (5.29) with respect to μ, we find that d is proper if and only if

$$\int_\Omega V^{-\frac{1}{2}}(y) \exp\left\{-\frac{\lambda}{2}d(y;\mu)\right\} \frac{\partial d}{\partial \mu}(y;\mu)\, dy = 0 \qquad (5.32)$$

for all $\mu \in \Omega$ and $\lambda \in \Lambda$. There seem to be no obvious analytical methods available for constructing solutions to this equation. However, the equation may be useful for numerical work, and for a particular unit deviance d, we may proceed by first checking numerically that (5.32) holds, or that (5.29) does not depend on μ, before proceeding to an analytical proof that d is proper.

Unfortunately, the class of proper unit deviances is fairly restricted compared with the class of all regular unit deviances. The scarcity of proper unit deviances is illustrated by the generalized inverse Gaussian distribution in Section 5.3.1 below, where only three out of an infinite class of unit deviances are proper. Thus, while the above method works in principle, and many of the examples of Section 5.3 were found by this method, it may not always lead to good results in practice.

5.2.2 Transformational dispersion models

We now consider a class of models where the above method of construction of proper dispersion models works well. Recall from Chapter 1 that a **location-dispersion model** is defined by

$$p(y;\mu,\lambda) = c(\lambda) \exp\left\{-\frac{\lambda}{2}d(y - \mu)\right\},$$

for $y, \mu \in \mathbf{R}$, where $d(\,\cdot\,)$ is non-negative and satisfies $d(0) = 0$. Any such d may be used as long as the exponential factor of the density is integrable. In particular, if the unit deviance $d(y - \mu)$ is regular, the corresponding unit variance function is constant, and the model is a regular proper dispersion model. We now generalize this idea.

Let G be a group acting on the interval Ω, the action being written as

$$(g, y) \mapsto gy \quad \text{for} \quad (g, y) \in G \times \Omega.$$

Let t be an arbitrary function on Ω, and let b be a function on Ω that is invariant under the action of G, i.e. $b(gy) = b(y)$ for all g and y. Suppose that the integral

$$\frac{1}{c(\lambda)} = \int_\Omega b(y) \exp\left\{\lambda t\left(g^{-1}y\right)\right\} dy \qquad (5.33)$$

CONSTRUCTION OF PROPER DISPERSION MODELS   187

is finite for λ in some interval $\Lambda \subseteq \mathbf{R}_+$ unbounded to the right. Then the integral depends on (g, λ) only through λ, as indicated in the notation $c(\lambda)$, due to the invariance of b. Hence, following Jørgensen (1983), we obtain a general proper dispersion model,

$$p(y; g, \lambda) = c(\lambda) b(y) \exp\left\{\lambda t \left(g^{-1} y\right)\right\}, \quad y \in \Omega. \tag{5.34}$$

We call (5.34) a **transformational dispersion model**. In the terminology of Barndorff-Nielsen, Blæsild, Jensen and Jørgensen (1982) the model (5.34) is an example of a composite transformation model with index parameter λ.

Let us now assume that G acts freely and transitively on Ω, i.e. Ω and G are in one-to-one correspondence. Then, assuming that the supremum

$$\hat{t} = \hat{t}(y) = \sup_{g \in G} t\left(g^{-1} y\right)$$

is finite, it does not depend on y. Note that the assumption of a transitive action is essential here. The one-to-one correspondence between g and y implies that $t\left(g^{-1} y\right)$ is yokable, and the corresponding unit deviance d is defined by

$$d(y; \mu) = 2\left\{\hat{t} - t\left(\hat{g}_\mu^{-1} y\right)\right\},$$

where $\hat{g}_\mu$ is the maximum likelihood estimate of g corresponding to the observation $\mu \in \Omega$. We then obtain a general proper dispersion model with standard representation

$$p(y; \hat{g}_\mu, \lambda) = c(\lambda) e^{\lambda \hat{t}} \widehat{b(y)} \exp\left\{-\frac{\lambda}{2} d(y; \mu)\right\}, \quad y \in \Omega. \tag{5.35}$$

If furthermore d is regular, Corollary 5.5 implies that Barndorff-Nielsen's formula is exact for (5.34), due to the constancy of $\hat{t}$. The model is then a regular proper dispersion model with $b(y) \propto V^{-1/2}(y)$, where V is the unit variance function corresponding to d.

The condition that $t\left(g^{-1} y\right)$ be yokable is closely associated with the action of G on Ω being free, because yokability implies that int G, suitably defined, is in one-to-one correspondence with Ω. However, there are many cases where t is yokable without the action being free.

In the special case where G is the group of translations of $\mathbf{R}$, the transformational dispersion model (5.35) becomes a location-dispersion model, as considered in Chapter 1. Similarly, the multiplicative action on $\mathbf{R}_+$ gives rise to the class of scale-dispersion models. The von Mises distribution is a transformational dispersion model corresponding to the action $y \mapsto (g+y) \mod 2\pi$, where G is the group of rotations of the unit circle (the orthogonal group).

The assumption in the above construction that Ω be an interval is not crucial, and the construction remains valid if Ω is a discrete set. However, as a rule, transformation models for discrete data are not very useful because the parameter space is then also discrete. An exception are the ranking models introduced by Mallows (1957), see also Critchlow (1985). Mallows' (1957) models have the form

$$p(y;\mu,\lambda) = c(\lambda)\exp\left\{-\frac{\lambda}{2}d(y;\mu)\right\},$$

where y and μ are elements of the permutation group S_n, the set of permutations of the set $\{1,\ldots n\}$, and where d is a metric on S_n. The interpretation of this model is similar to that of dispersion models. Thus, μ may be interpreted as the most likely permutation in S_n, and the probability of a given ranking y decreases exponentially according to the distance $d(y;\mu)$ from y to μ. McCullagh (1993) observed that certain metrics d may be interpreted in terms of Euclidean distance, in which case there is a close analogy between the Mallows model and the von Mises–Fisher model on the sphere.

5.2.3 Reproductive exponential families

Daniels (1980) showed the following remarkable result, expressed here in our terminology.

Theorem 5.6 (Daniels) *The only exponential dispersion models with exact renormalized saddlepoint approximations are the normal, gamma and inverse Gaussian distributions.*

For a proof, see Daniels (1980) or Blæsild and Jensen (1985). This result implies that the intersection between the classes of exponential and regular proper dispersion models consists of precisely the above three models, as we shall now see. This was illustrated in the 'world map' of dispersion models in Chapter 1, Figure 1.2.

Consider a steep continuous exponential dispersion model $Y \sim \text{ED}(\mu,\sigma^2)$ with canonical parameter domain Θ. Due to the steepness, the support is Ω, and the density takes the following form, for $y \in \Omega$:

$$\begin{aligned}p(y;\theta,\lambda) &= c(y;\lambda)\exp\left[\lambda\{y\theta - \kappa(\theta)\}\right] \quad (5.36)\\ &= c(y;\lambda)\exp\left\{\lambda\hat{t}(y) - \frac{\lambda}{2}d(y;\mu)\right\},\end{aligned}$$

where the last expression gives the standard form of the density. The unit deviance is

$$d(y;\mu) = -2\left[y\tau^{-1}(\mu) - \kappa\left\{\tau^{-1}(\mu)\right\} - \hat{t}(y)\right], \quad (5.37)$$

where

$$\hat{t}(y) = y\tau^{-1}(y) - \kappa\left\{\tau^{-1}(y)\right\}.$$

The corresponding renormalized saddlepoint approximation takes the usual form

$$p_0(y;\mu,\lambda) = c_0(\mu,\lambda)V^{-\frac{1}{2}}(y)\exp\left\{-\frac{\lambda}{2}d(y;\mu)\right\}, \quad (5.38)$$

where $c_0(\mu,\lambda)$ is a normalizing constant and V is the unit variance function for d. Now, assume that the renormalized saddlepoint approximation is exact, so that (5.36) and (5.38) are identical. Then their moment generating functions are also identical, and this implies that for all u such that $\theta + u/\lambda \in \Theta$,

$$\exp\left[\lambda\left\{\kappa\left(\theta + \frac{u}{\lambda}\right) - \kappa(\theta)\right\}\right]$$
$$= \exp\left[\lambda\left\{\kappa\left(\theta + \frac{u}{\lambda}\right) - \kappa(\theta)\right\}\right]\frac{c_0\left\{\tau(\theta),\lambda\right\}}{c_0\left\{\tau\left(\theta + \frac{u}{\lambda}\right),\lambda\right\}}.$$

Hence, $c_0(\mu,\lambda)$ depends on (μ,λ) only through λ. Consequently the unit deviance d is proper, and (5.38) becomes a regular proper dispersion model,

$$p(y;\mu,\lambda) = c(\lambda)V^{-\frac{1}{2}}(y)\exp\left\{-\frac{\lambda}{2}d(y;\mu)\right\}, \quad (5.39)$$

say.

Note that the renormalized saddlepoint approximation is exact for all regular proper dispersion models. It hence follows from Daniels' theorem above that the only dispersion models that are both exponential and regular proper are the three mentioned in the theorem.

As we saw in Chapter 1, the normal distribution is a location-dispersion model, and the gamma distribution is a scale-dispersion model. Hence, both are transformational dispersion models. The inverse Gaussian distribution is closed under scale transformations, as we saw in Chapter 4, but it is not transformational in the above sense, because both μ and σ^2 are affected by scale transformations in this case.

Due to the particular form of the unit deviance (5.37), the family (5.39) is a reproductive exponential family in the sense of Barndorff-Nielsen and Blæsild (1983, 1988). More generally, both (5.39) and the renormalized saddlepoint approximation (5.38) are two-parameter exponential families with canonical statistic $\{y, \hat{t}(y)\}$.

5.2.4 Non-transformational dispersion models

While we have straightforward methods for constructing exponential and transformational dispersion models, no such method of comparable generality exists for proper dispersion models that are not transformational. In particular, the method based on identifying proper unit deviances, when applied outside the class of transformational dispersion models, is a 'trial and error' method, with low success probability.

Yet, there are many examples of regular proper dispersion models that are not of the transformational form, as we shall see in Section 5.3, making the distinction between transformational and non-transformational dispersion models crucial. These two classes of dispersion models were represented on the 'world map' of dispersion models (Figure 1.2) as the northern and southern hemispheres, respectively.

5.3 Examples

We now consider a number of examples of dispersion models.

5.3.1 Generalized inverse Gaussian distribution

In Section 4.4.4, we found the natural exponential families in the generalized inverse Gaussian distribution obtained by fixing two parameters, from which we, in turn, generated exponential dispersion models. We shall now identify subclasses of the generalized inverse Gaussian distribution that are proper dispersion models.

Fixed γ

Recall from Section 4.4.4 that the generalized inverse Gaussian distribution $\mathrm{GIG}(\gamma, \chi, \psi)$ has probability density function

$$p(y; \gamma, \chi, \psi) = \frac{(\psi/\chi)^{\gamma/2}}{2K_\gamma\left(\sqrt{\chi\psi}\right)} y^{\gamma-1} \exp\left\{-\frac{1}{2}(\chi y^{-1} + \psi y)\right\}, \quad (5.40)$$

for $y > 0$. For given $\chi, \psi > 0$ and $\gamma \in \mathbf{R}$, let $\chi = \lambda \mu^{2\gamma+1}$ and $\psi = \lambda \mu^{2\gamma-1}$ in (5.40), where λ and μ are positive. For γ known,

EXAMPLES   191

we then obtain a renormalized saddlepoint approximation,

$$c_\gamma(\mu,\lambda)y^{\gamma-1}\exp\left\{-\frac{\lambda(y-\mu)^2}{2y\mu^{1-2\gamma}}\right\}. \quad (5.41)$$

The unit deviance is

$$d_\gamma(y;\mu) = \frac{(y-\mu)^2}{y\mu^{1-2\gamma}}$$

with corresponding unit variance function $V_\gamma(\mu) = \mu^{2(1-\gamma)}$. The normalizing constant $c_\gamma(\mu,\lambda)$ of (5.41) is defined by

$$\frac{1}{c_\gamma(\mu,\lambda)} = 2K_\gamma(\lambda\mu^{2\gamma})e^{\lambda\mu^{2\gamma}}\mu^\gamma. \quad (5.42)$$

By an asymptotic expansion for K_γ (Abramowitz and Stegun, 1972, p. 378) we obtain the following asymptotic expansion of $1/c_\gamma(\mu,\lambda)$ for λ large:

$$\left\{\frac{2\pi}{\lambda}\right\}^{\frac{1}{2}}\left\{1 + \frac{4\gamma^2-1}{8\lambda\mu^{2\gamma}} + \frac{(4\gamma^2-1)(4\gamma^2-9)}{2!(8\lambda\mu^{2\gamma})^2}\right.$$
$$\left. + \frac{(4\gamma^2-1)(4\gamma^2-9)(4\gamma^2-25)}{3!(8\lambda\mu^{2\gamma})^3} + \cdots\right\}. \quad (5.43)$$

It follows that only the three values $\gamma = \pm 1/2$ and $\gamma = 0$ make $c_\gamma(\mu,\lambda)$ independent of μ, in turn making the unit deviance d_γ proper. The three corresponding regular proper dispersion models are also exponential family dispersion models. We now consider each of these three cases in more detail.

In the cases $\gamma = \pm 1/2$ we have

$$c_{\pm 1/2}(\mu,\lambda) = \sqrt{\frac{\lambda}{2\pi}}.$$

For $\gamma = -1/2$ we obtain the inverse Gaussian distribution, which we have already studied in Chapter 4. For $\gamma = 1/2$ we obtain the distribution of the reciprocal of an inverse Gaussian variate, denoted RIG(μ,σ^2), with probability density function for $y > 0$ given by

$$p(y;\mu,\lambda) = \sqrt{\frac{\lambda}{2\pi}}y^{-\frac{1}{2}}\exp\left\{-\frac{\lambda(y-\mu)^2}{2y}\right\}. \quad (5.44)$$

This distribution is known as the **reciprocal inverse Gaussian distribution**.

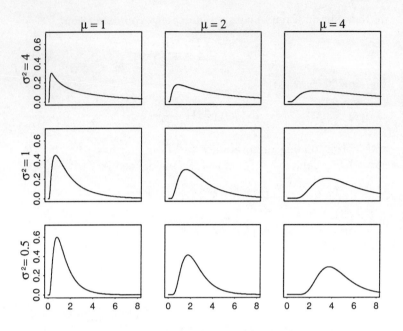

Figure 5.1. *Reciprocal inverse Gaussian densities.*

Figure 5.1 shows some examples of reciprocal inverse Gaussian densities. The unit variance function of the reciprocal inverse Gaussian distribution is $V(\mu) = \mu$, giving an example of a continuous dispersion model with the same unit variance function as the Poisson distribution.

The third case where a proper dispersion model is obtained corresponds to $\gamma = 0$, where (5.41) turns into

$$p(y; \mu, \lambda) = \frac{e^{-\lambda}}{2K_0(\lambda)} y^{-1} \exp\left\{ -\frac{\lambda(y-\mu)^2}{2y\mu} \right\}. \tag{5.45}$$

This distribution is known as the **hyperbola distribution** (Barndorff-Nielsen, 1978b; Jensen, 1981), see Exercise 1.8. The corresponding unit variance function is $V(\mu) = \mu^2$.

Fixed α

We now consider a different parametrization of the generalized inverse Gaussian distribution, by which the family may be divided into proper dispersion models.

EXAMPLES

Following Jørgensen (1982), we define the new parameters

$$\eta = \sqrt{\frac{\chi}{\psi}}, \ \omega = \sqrt{\chi\psi}.$$

Note that η is a scale parameter and ω is invariant under scale transformations. Using the parametrization (γ, η, ω), the density (5.40) takes the form

$$\frac{y^{\gamma-1}}{2K_\gamma(\omega)} \exp\left\{-\gamma \log \eta - \frac{\omega}{2}\left(\frac{y}{\eta} + \frac{\eta}{y}\right)\right\}. \quad (5.46)$$

In spite of a certain resemblance, this is not a dispersion model for fixed γ, except in the special case $\gamma = 0$ which gives the hyperbola distribution with position η and dispersion $1/\omega$.

To obtain a dispersion model for other values of γ with dispersion parameter $1/\omega$, we need to make ω a common factor in the exponent of (5.46). We are hence led to replace the parameter γ by $\alpha = \gamma/\omega$, in terms of which the density becomes

$$\frac{y^{-1}}{2K_{\alpha\omega}(\omega)} \exp\left[-\omega\left\{\alpha \log \frac{\eta}{y} + \frac{1}{2}\left(\frac{y}{\eta} + \frac{\eta}{y}\right)\right\}\right].$$

This parametrization gives us an exponential family dispersion model for each known value of α.

To obtain a density of standard form, we consider estimation of η. By maximizing the function

$$t_\alpha(y; \eta) = \alpha \log \frac{y}{\eta} - \frac{1}{2}\left(\frac{y}{\eta} + \frac{\eta}{y}\right),$$

we obtain the following estimate of η for ω and α known:

$$\eta = y\left(\sqrt{\alpha^2 + 1} - \alpha\right).$$

This shows that t_α is yokable, and we hence replace η by the new parameter μ defined by

$$\mu = \frac{\eta}{\sqrt{\alpha^2 + 1} - \alpha}.$$

Writing $\bar{\alpha} = \sqrt{\alpha^2 + 1}$, the corresponding unit deviance is

$$2\alpha \log \frac{\mu}{y} + \frac{y}{\mu}(\bar{\alpha} + \alpha) + \frac{\mu}{y}(\bar{\alpha} - \alpha) - 2\bar{\alpha}. \quad (5.47)$$

The parametrization (α, μ, ω) gives us a regular proper dispersion model for each fixed value of α with location μ and dispersion $1/\omega$.

The parametrization (α, μ, ω) does not take proper account of the gamma and reciprocal gamma cases, and we hence make a further reparametrization to remedy this problem.

The solution is to make the unit deviance include the gamma and reciprocal gamma unit deviances as limiting cases. This leads us to reparametrize from ω to the new index parameter

$$\lambda = \omega\sqrt{\alpha^2 + 1}. \tag{5.48}$$

The corresponding unit deviance d_β is defined by dividing (5.47) by $\sqrt{\alpha^2 + 1}$, giving

$$d_\beta(y; \mu) = 2\beta \log \frac{\mu}{y} + \frac{y}{\mu}(1+\beta) + \frac{\mu}{y}(1-\beta) - 2,$$

where $\beta = \alpha/\sqrt{\alpha^2 + 1}$. It is also convenient to reparametrize from α to β, making the gamma and reciprocal gamma cases correspond to finite values of β, namely $\beta = \pm 1$, respectively. The hyperbola distribution corresponds to $\beta = 0$. Note that neither the inverse Gaussian distribution nor its reciprocal correspond to a fixed value of β.

The domain of the new parameter (β, μ, λ) is given by $\beta \in [-1, 1]$ and $\mu, \lambda > 0$. The corresponding version of the density is

$$\frac{\left(\frac{1+\beta}{1-\beta}\right)^{\frac{\lambda\beta}{2}} e^{-\frac{\lambda}{2}}}{2y K_{\lambda\beta}\left(\lambda\sqrt{1-\beta^2}\right)} \exp\left\{-\frac{\lambda}{2} d_\beta(y; \mu)\right\}. \tag{5.49}$$

For each fixed value of $\beta \in [-1, 1]$ we obtain a regular proper dispersion model, in fact a scale-dispersion model, and also an exponential family dispersion model.

Each of these proper dispersion models has the same unit variance function, $V(\mu) = \mu^2$. This provides a vivid illustration of the phenomenon that, contrary to the exponential dispersion model case, a proper dispersion model is not characterized by its unit variance function. Still, the unit variance function determines the second-order behaviour of the unit deviance near its minimum, so one would not expect major variations in shape for different values of the parameter β, in particular when λ is large. This is confirmed by the plots in Figure 5.2, which show some generalized inverse Gaussian densities of the form (5.49), all with $\mu = 1$, for various values of σ^2 and β.

It is probably rare to identify more than one type of proper dispersion models within a three-parameter family, as we did here for the generalized inverse Gaussian distribution.

EXAMPLES 195

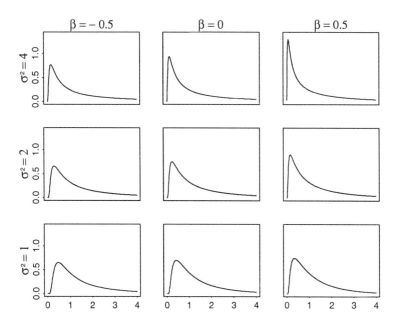

Figure 5.2. *Generalized inverse Gaussian densities with $\mu = 1$.*

5.3.2 Leipnik's distribution

Let us consider the distribution given by the following probability density function:

$$\frac{\left(1-x^2\right)^{\frac{\lambda-1}{2}}}{B\left(\frac{\lambda+1}{2}, \frac{1}{2}\right)} \left(1 - 2x\theta + \theta^2\right)^{-\frac{\lambda}{2}}$$

$$= \frac{\left(1-x^2\right)^{-\frac{1}{2}}}{B\left(\frac{\lambda+1}{2}, \frac{1}{2}\right)} \exp\left\{-\frac{\lambda}{2} \log\left(\frac{1 - 2x\theta + \theta^2}{1 - x^2}\right)\right\}, \quad (5.50)$$

for $x, \theta \in (-1, 1)$ and $\lambda > -1$, where B denotes the beta function. Leipnik (1947) derived this as a 'smoothed' approximation to the distribution for the circular serial correlation coefficient for a sample of size λ. Its properties were subsequently studied by several authors, including Jenkins (1956), White (1957) and Kemp (1970). McCullagh (1989) rediscovered the distribution as a kind of noncentral version of the symmetric beta family, and noted a connection with Brownian motion. McCullagh (1989) also studied statistical inference for the two parameters θ and $\gamma = \lambda/2$.

For further details, see Johnson, Kotz and Balakrishnan (1995, pp. 612–617). We refer to (5.50) as **Leipnik's distribution**.

From our point of view, (5.50) is a regular proper dispersion model with position parameter θ and unit variance function $V(\theta) = \left(1 - \theta^2\right)/2$ defined on $\Omega = (-1, 1)$. McCullagh's (1989) analysis of this distribution was seminal for the development of proper dispersion models, because many of the properties derived by McCullagh have direct generalizations to the whole class of proper dispersion models. McCullagh (1989) also introduced a second distribution somewhat similar to (5.50), but this distribution is not a dispersion model, see Exercise 5.3.

A crucial property of Leipnik's distribution is that the domain for the index parameter λ contains negative values, considerably complicating the statistical analysis. The likelihood has a singularity for $\lambda = 0$, where it is constant as a function of the position θ. For $\lambda < 0$, the usual interpretations of θ and λ break down, because of the change of sign of the multiplier of the unit deviance in the exponent of the density (5.50). However, so far this feature is unique for Leipnik's distribution, because we have found no other examples of dispersion models where negative values of the index parameter occur.

It is useful to transform the distribution to the unit interval and reparametrize,

$$x \mapsto y = (x+1)/2, \quad \theta \mapsto \mu = (\theta+1)/2.$$

This gives the following form of the density:

$$p(y; \mu, \lambda) = \frac{\{y(1-y)\}^{-\frac{1}{2}}}{B\left(\frac{\lambda+1}{2}, \frac{1}{2}\right)} \exp\left\{-\frac{\lambda}{2} d(y; \mu)\right\}, \qquad (5.51)$$

for $y, \mu \in (0, 1)$ and $\lambda > -1$, where the unit deviance d is defined by

$$d(y; \mu) = \log\left\{1 + \frac{(y-\mu)^2}{y(1-y)}\right\}.$$

The corresponding unit variance function is $V(\mu) = \mu(1 - \mu)$, the same as for the binomial distribution. Note that $\lambda = 0$ gives the beta distribution with parameters $(1/2, 1/2)$. We refer to (5.51) as the **transformed Leipnik distribution**. For later use, note that the unit deviance $d(y; \mu)$ is a one-to-one function of the unit deviance $(y - \mu)^2 / \{y(1-y)\}$, so that, by Lemma 5.2, the latter is proper.

EXAMPLES

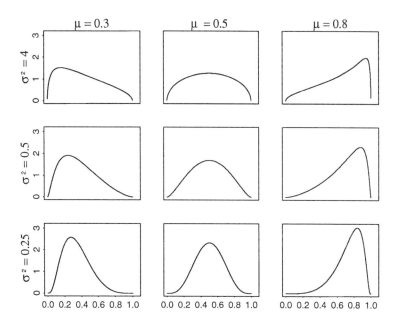

Figure 5.3. *Some densities of the transformed Leipnik distribution.*

Figure 5.3 shows some densities of the transformed Leipnik distribution, indexed by $\sigma^2 = 1/\lambda$. The densities are considerably more dispersed than the corresponding simplex densities examined in Chapter 1, although this is mainly due to the different unit variance functions.

Bondesson class

Bondesson (1990) considered a class of distributions with densities of the form

$$a_0\, x^{\beta-1} \prod_{i=1}^{k}(1 + c_i x)^{-\lambda_i}, \tag{5.52}$$

for $x > 0$, where a_0 is a normalizing constant, and β, $\lambda_1, \ldots, \lambda_k$ and $c_1, \ldots, c_k$ are positive parameters. Bondesson's (1990) purpose was the study of infinite divisibility, but there is a curious relation between the transformed Leipnik distribution and (5.52). Making the transformation $y = x/(x+1)$ and taking $c_1 = 1$, the density of

y becomes

$$a_1 y^{\alpha_1-1}(1-y)^{\alpha_2-1}\prod_{i=2}^{k}\left(\frac{1}{y}+c_i\frac{1}{1-y}\right)^{-\lambda_i}$$
$$= a_2 y^{\alpha_1-1}(1-y)^{\alpha_2-1}\exp\left[\sum_{i=2}^{k}\lambda_i\log\left\{1+\frac{(y-\mu_i)^2}{y(1-y)}\right\}\right],$$

for $y \in (0,1)$, where a_1 and a_2 are new normalizing constants. The new parameters are defined by $\alpha_1 = \beta - \lambda_2 - \cdots - \lambda_m \in \mathbf{R}$, $\alpha_2 = \lambda_1 - \beta \in \mathbf{R}$, and $\mu_i = 1/\left(1+\sqrt{c_i}\right) \in (0,1)$ for $i = 2,\ldots,k$. The special case $k=2$, $\alpha_1 = \alpha_2 = 1/2$ corresponds to the transformed Leipnik distribution.

5.3.3 Simplex distributions

Proper dispersion models on the unit interval, such as the transformed Leipnik distribution, complement exponential dispersion models in an essential way, because there are no infinitely divisible exponential dispersion models on the unit interval. Without infinite divisibility, inference on λ is considerably complicated, whereas infinite divisibility is not a concern for proper dispersion models. We now extend the simplex distribution from Chapter 1 to a wider class of distributions on the unit interval.

Definition

Recall that the simplex distribution $S^-(\mu, \sigma^2)$ with parameters $\mu \in (0,1)$ and $\sigma^2 = 1/\lambda > 0$ is defined by the probability density function

$$p(y;\mu,\lambda) = \sqrt{\frac{\lambda}{2\pi\{y(1-y)\}^3}}\exp\left\{-\frac{1}{2\sigma^2}d(y;\mu)\right\}, \qquad (5.53)$$

for $0 < y < 1$, where the unit deviance d is defined by

$$d(y;\mu) = \frac{(y-\mu)^2}{y(1-y)\mu^2(1-\mu)^2}. \qquad (5.54)$$

Now, consider the class of renormalized saddlepoint approximations defined by

$$c(\alpha_1,\alpha_2,\mu,\lambda)y^{\alpha_1-1}(1-y)^{\alpha_2-1}\exp\left\{-\frac{\lambda}{2}d_{\alpha_1,\alpha_2}(y;\mu)\right\}, \qquad (5.55)$$

EXAMPLES

where $d_{\alpha_1,\alpha_2}(y;\mu)$ is a unit deviance defined by

$$d_{\alpha_1,\alpha_2}(y;\mu) = \mu^{2\alpha_1-1}(1-\mu)^{2\alpha_2-1}\frac{(y-\mu)^2}{y(1-y)}, \qquad (5.56)$$

the corresponding unit variance function being

$$V_{\alpha_1,\alpha_2}(\mu) = \mu^{2(1-\alpha_1)}(1-\mu)^{2(1-\alpha_2)}.$$

The parameters of this distribution are $(\alpha_1,\alpha_2)^\top \in \mathbf{R}^2$, $\sigma^2 = 1/\lambda > 0$ and $\mu \in (0,1)$. The distributions (5.55) are called **simplex distributions**, and are denoted $Y \sim S(\alpha_1,\alpha_2,\mu,\sigma^2)$. The special case $(\alpha_1,\alpha_2) = (-1/2,-1/2)$ gives the original simplex distribution $S^-(\mu,\sigma^2)$, which we from now on refer to as the **standard simplex distribution**. If $\alpha_1,\alpha_2 > 0$, the limiting case $\lambda = 0$ is the beta distribution with parameters (α_1,α_2).

Properties

The family (5.55) is a four-parameter exponential family. One version of the canonical statistic is

$$\left[\frac{Y}{1-Y}, \frac{1-Y}{Y}, \log\frac{Y}{1-Y}, \log\{Y(1-Y)\}\right]^\top$$

which includes the odds and log odds ratios. This provides an intuitive justification for the particular exponential family (5.55) for data concerning proportions. The transformation $Y \mapsto 1-Y$ transforms $S(\alpha_1,\alpha_2,\mu,\sigma^2)$ into $S(\alpha_2,\alpha_1,1-\mu,\sigma^2)$.

The density (5.55) may have one or two modes, and when $\lambda > 0$, the density tends to zero at $y = 0,1$. If $\alpha_1 + \alpha_2 \geq 2$ there is only one mode, whereas for $\alpha_1 + \alpha_2 < 2$, there may be two modes if λ is small. These conclusions follow by noting that the stationary points of the log density are determined by a cubic equation.

The mixed moments of Y and $1-Y$ may be expressed in terms of the normalizing constant $c(\alpha_1,\alpha_2,\mu,\lambda)$. Thus,

$$EY^\alpha(1-Y)^\beta = \frac{c(\alpha_1,\alpha_2,\mu,\lambda)}{c\left\{\alpha_1+\alpha,\alpha_2+\beta,\mu,\frac{\lambda}{\mu^{2\alpha}(1-\mu)^{2\beta}}\right\}}. \qquad (5.57)$$

From (5.57) we may calculate EY and $EY(1-Y)$, and hence the mean and variance. Using results from Table 5.1 we obtain for $(\alpha_1,\alpha_2) = (-1/2,-1/2)$ that $EY = \mu$ and $\mathrm{var}\, Y$ is given by

$$\mu(1-\mu) - \sqrt{\frac{\lambda}{2}}\exp\left\{\frac{\lambda}{\mu^2(1-\mu)^2}\right\}\Gamma\left\{\frac{1}{2},\frac{\lambda}{2\mu^2(1-\mu)^2}\right\}, \qquad (5.58)$$

Table 5.1. *Simplex distributions*

(α_1, α_2)	$1/c(\lambda)$	$V(y)$	$d(y;\mu)$
$(\frac{1}{2}, \frac{1}{2})$	$\sqrt{\pi e^\lambda}\Gamma(\frac{1}{2}, \frac{\lambda}{2})$	$y(1-y)$	$\frac{(y-\mu)^2}{y(1-y)}$
$(0, 0)$	$2e^\lambda K_0(\lambda)$	$y^2(1-y)^2$	$\frac{(y-\mu)^2}{y(1-y)\mu(1-\mu)}$
$(-\frac{1}{2}, -\frac{1}{2})$	$\sqrt{2\pi/\lambda}$	$y^3(1-y)^3$	$\frac{(y-\mu)^2}{y(1-y)\mu^2(1-\mu)^2}$
$(-\frac{1}{2}, \frac{1}{2})$	$\sqrt{2\pi/\lambda}$	$y^3(1-y)$	$\frac{(y-\mu)^2}{y(1-y)\mu^2}$
$(\frac{1}{2}, -\frac{1}{2})$	$\sqrt{2\pi/\lambda}$	$y(1-y)^3$	$\frac{(y-\mu)^2}{y(1-y)(1-\mu)^2}$

in terms of the incomplete gamma function defined by

$$\Gamma(a, x) = \int_x^\infty t^{a-1} e^{-t} \, dt. \tag{5.59}$$

Special cases

Table 5.1 summarizes five cases where a regular proper dispersion model is obtained in (5.55) by fixing (α_1, α_2). The second column gives the reciprocal of $c(\lambda) = c(\alpha_1, \alpha_2, \mu, \lambda)$. These five models are all exponential family dispersion models. We now show how to calculate the normalizing constant $c(\alpha_1, \alpha_2, \mu, \lambda)$ in these cases.

Consider the transformation $x = y/(1-y)$, the inverse of the transformation used in connection with the Bondesson class, which transforms $(0, 1)$ to $\mathbf{R}_+$. Letting $\xi = \mu/(1-\mu)$ and applying the transformation to (5.55), we obtain the following density:

$$c(\alpha_1, \alpha_2, \mu, \lambda) \frac{x^{\alpha_1-1}}{(1+x)^{\alpha_1+\alpha_2}} \exp\left\{-\frac{\lambda}{2} \bar{d}_{\alpha_1, \alpha_2}(x; \xi)\right\}, \tag{5.60}$$

where

$$\bar{d}_{\alpha_1, \alpha_2}(x; \xi) = \frac{\xi^{2\alpha_1-1}(x-\xi)^2}{x(1+\xi)^{2(\alpha_1+\alpha_2)}}.$$

If $\alpha_1 + \alpha_2$ is zero, we find that (5.60) is identical to the generalized inverse Gaussian distribution (5.41). In this case, the simplex distribution is hence a simple transformation of the generalized inverse Gaussian distribution. This accounts for three of the cases in Table 5.1.

If $\alpha_1 + \alpha_2$ is a negative integer, we make a binomial expansion for $(1+x)^{-\alpha_1-\alpha_2}$ and express $c(\alpha_1, \alpha_2, \mu, \lambda)$ in terms of the normalizing constant for the generalized inverse Gaussian distribution

EXAMPLES

from equation (5.42). This, and the formula

$$K_{\pm\frac{1}{2}}(\lambda) = \sqrt{\frac{\pi}{2\lambda}}e^{-\lambda},$$

account for the case $(\alpha_1, \alpha_2) = (-1/2, -1/2)$ in Table 5.1.

In the case $(\alpha_1, \alpha_2) = (1/2, 1/2)$, the density (5.55) takes the form

$$c\left(\frac{1}{2}, \frac{1}{2}, \mu, \lambda\right) \{y(1-y)\}^{-1/2} \exp\left\{-\frac{\lambda}{2}\frac{(y-\mu)^2}{y(1-y)}\right\}. \quad (5.61)$$

As noted in Section 5.3.2 above, the corresponding unit deviance is proper. We may hence take $\mu = 0$ and calculate $f = 1/c$ as follows:

$$\begin{aligned} f(\lambda) &= \int_0^1 \{y(1-y)\}^{-1/2} \exp\left\{-\frac{\lambda y}{2(1-y)}\right\} dy \\ &= \int_0^\infty (1+x)^{-1} x^{-1/2} e^{-\lambda x/2} \, dx. \end{aligned}$$

By differentiating under the integral sign we find that f satisfies the differential equation

$$f'(\lambda) - \frac{1}{2}f(\lambda) = -\sqrt{\frac{\pi}{2\lambda}}.$$

Hence

$$f(\lambda) = \sqrt{\pi e^\lambda}\,\Gamma\left(\frac{1}{2}, \frac{\lambda}{2}\right),$$

where $\Gamma(\cdot\,,\cdot)$ is the incomplete gamma function (5.59).

Figures 5.4 and 5.5 show density functions for two of the cases in Table 5.1, illustrating in particular the case of a bimodal density. Together with the corresponding plots for the standard simplex distribution in Chapter 1, these plots show that the simplex distributions can take on a considerable variety of shapes.

Based on the apparent analogy between the simplex distribution and the generalized inverse Gaussian distribution (5.41), and confirmed by a small numerical investigation, it seems unlikely that there are other values of (α_1, α_2) for which we obtain proper dispersion models in (5.55).

5.3.4 Simplex-binomial mixtures

We now consider briefly a simplex-binomial mixture model for binomial data. It is similar to the beta-binomial distribution, but uses the simplex instead of the beta as mixing distribution.

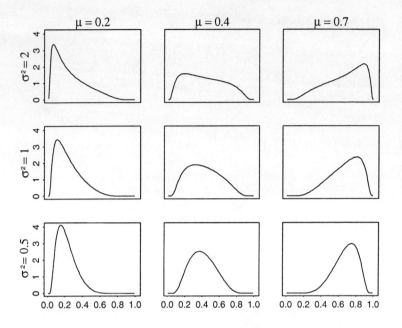

Figure 5.4. *Simplex distributions for* $(\alpha_1, \alpha_2) = (0, 0)$.

The new distribution is defined as the following binomial mixture:
$$Y|P = p \sim \text{Bi}(m, p)$$
$$P \sim S(\alpha_1, \alpha_2, \mu, \sigma^2),$$
with support $\{0, \ldots, m\}$, where $\sigma^2 = 1/\lambda > 0$ and $\mu \in (0, 1)$. It is useful to write the simplex density as follows:
$$\frac{\bar{a}_0(\alpha_1, \alpha_2, \mu, \sigma)}{\sqrt{2\pi\sigma^2}} p^{\alpha_1 - 1}(1 - p)^{\alpha_2 - 1} \exp\left\{\lambda t_{\alpha_1, \alpha_2}(p; \mu)\right\},$$
where
$$t_{\alpha_1, \alpha_2}(p; \mu) = -\frac{(p - \mu)^2}{2p(1 - p)} \mu^{2\alpha_1 - 1}(1 - \mu)^{2\alpha_2 - 1}.$$
Note the special form of the normalizing constant which simplifies some formulas in the following. Note also that $\bar{a}_0$ is asymptotically 1 for σ^2 small, by the saddlepoint approximation.

The distribution of Y is given by the probability function
$$\frac{\binom{m}{y} \bar{a}_0(\alpha_1, \alpha_2, \mu, \sigma)}{\sqrt{2\pi\sigma^2}} \int_0^1 p^{\alpha_1 + y - 1}(1 - p)^{\alpha_2 + m - y - 1} e^{\lambda t_\alpha(p; \mu)} \, dp.$$

EXAMPLES

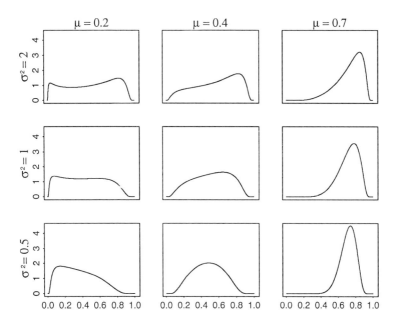

Figure 5.5. *Simplex distributions for* $(\alpha_1, \alpha_2) = (0.5, -0.5)$.

If we write $t_{\alpha_1,\alpha_2}(p;\mu)$ as follows:

$$t_{\alpha_1,\alpha_2}(p;\mu) = \mu^{-2y}(1-\mu)^{-2(m-y)} t_{\alpha_1+y,\alpha_2+m-y}(p;\mu),$$

the integral is given in terms of the simplex normalizing constant. The probability function hence takes the form

$$\binom{m}{y}\mu^y(1-\mu)^{m-y} \frac{\bar{a}_0(\alpha_1, \alpha_2, \mu, \sigma)}{\bar{a}_0\{\alpha_1+y, \alpha_2+m-y, \mu, \sigma\mu^y(1-\mu)^{m-y}\}}, \quad (5.62)$$

which is the binomial probability function times a factor that tends to 1 as σ^2 goes to zero. There is no closed-form expression for the $\bar{a}_0$ function, which may have to be calculated by numerical integration. In the special case $\alpha_1, \alpha_2 = -1/2$, where $\bar{a}_0(1/2, 1/2, \mu, \sigma) = 1$, the probability function becomes

$$\frac{\binom{m}{y}\mu^y(1-\mu)^{m-y}}{\bar{a}_0\{y-\frac{1}{2}, m-y-\frac{1}{2}, \mu, \sigma\mu^y(1-\mu)^{m-y}\}}.$$

The simplex-binomial distributions provide a possible alternative to the beta-binomial for analysis of overdispersed binomial data. Being a generalization of the beta distribution, the simplex distribution is a conjugate prior for the binomial distribution, in the Bayesian sense.

5.3.5 Tweedie contractions

We now consider an example of a pseudo-dispersion model (see Definition 5.2), obtained as a conditional distribution based on Tweedie models.

Let X_1 and X_2 be independent with

$$X_i \sim \text{Tw}_p^*(\theta, \lambda_i), \quad i = 1, 2, \tag{5.63}$$

and let $W = X_1 + X_2$ denote their sum. Since W is sufficient for the parameter θ, the conditional distribution of X_1 given $W = x$ does not depend on the parameter θ, and is denoted

$$X_1 | X_1 + X_2 = x \sim \text{GTw}_p(\lambda_1, \lambda_2, x).$$

The corresponding conditional density function is

$$p(w|x; \lambda_1, \lambda_2) = \frac{c_p^*(w; \lambda_1) c_p^*(x - w; \lambda_2)}{c_p^*(x; \lambda_1 + \lambda_2)}, \tag{5.64}$$

where expressions for $c_p^*(w; \lambda)$ may be found in Section 4.2. We call the distribution $\text{GTw}_p(\lambda_1, \lambda_2, x)$ the **Tweedie contraction** (Jørgensen and Song, 1998). When $p \leq 0$ the contraction is concentrated on **R**. When $p \in (1, 2)$, the contraction is concentrated on the interval $[0, x]$ and for $p \geq 2$ on $(0, x)$. The case $p = 1$ is the binomial distribution $\text{Bi}\{x, \lambda_1/(\lambda_1 + \lambda_2)\}$ and $p = 2$, $x = 1$ is the beta distribution $\text{Beta}(\lambda_1, \lambda_2)$. The case $p = 3$, $x = 1$ is the standard simplex distribution.

The saddlepoint approximation provides a convenient approximation to $c_p^*(x; \lambda)$ for large values of the index parameter λ. Defining

$$g_p(x; \lambda) = x \tau_p^{-1}(x/\lambda) - \lambda \kappa_p \{\tau_p^{-1}(x/\lambda)\},$$

using the notation of Chapter 4, we obtain

$$c_p^*(x; \lambda) \sim \sqrt{\frac{\lambda^{p-1}}{2\pi x^p}} \exp\{-g_p(x; \lambda)\}, \tag{5.65}$$

for $x \in \Omega_p$.

Now, consider the case $p \in (1, 2) \cup (2, \infty)$. Then we obtain

$$g_p(x; \lambda) = q_p x^{2-p} \lambda^{p-1}, \tag{5.66}$$

where
$$q_p = \frac{1}{(p-1)(p-2)}.$$
If we furthermore assume $x = 1$ and parametrize the contraction as follows:
$$\text{GTw}_p\left(\frac{\gamma^{1/(p-1)}}{1-\mu}, \frac{\gamma^{1/(p-1)}}{\mu}, 1\right), \quad (5.67)$$
the saddlepoint approximation to the density (5.64) becomes
$$\sqrt{\frac{\gamma}{2\pi}} V^{-\frac{1}{2}}(w) \exp\left\{-\frac{\gamma}{2} d_p(w;\mu)\right\}, \quad (5.68)$$
corresponding to the unit deviance on $(0,1) \times (0,1)$ given by
$$2q_p \left[(1-\mu)^{1-p} w^{2-p} + \mu^{1-p}(1-w)^{2-p} - \{\mu(1-\mu)\}^{1-p}\right],$$
with unit variance function $V(\mu) = \{\mu(1-\mu)\}^p$.

In the case $p = 3$ the saddlepoint approximation is exact, and (5.68) is the probability density function of the standard simplex distribution. In the general case, the properties of the saddlepoint approximation imply that the contraction (5.67) is a pseudo-dispersion model. We hence have an example of a pseudo-dispersion model that is not a renormalized saddlepoint approximation.

5.4 Studentization

We now consider a case where the method of 'new proper dispersion models for old ones' provided by Lemma 5.2 may be interpreted in terms of a kind of Studentization of the model. This construction allows us to move back and forth between the given model and the Studentized model, the relationship between the two types being a generalization of that between the normal distribution and Student's t distribution.

5.4.1 From 'normal' to 'Student'

Let $\text{PD}(\mu, \sigma^2)$ with support Ω denote a regular proper dispersion model with unit deviance $d(y;\mu)$ and unit variance function V, referred to as the **null** model or the **normal** model. Recall that Lemma 5.2 says that if the unit deviance $d(y;\mu)$ is proper, then so is any transformed unit deviance $f\{d(y;\mu)\}$, where f is twice differentiable and satisfies $f(0) = 0$ and $f'(0) > 0$. In the following, we assume that $f'(0) = 1$, so that $f\{d(y;\mu)\}$ has the same unit variance function as $d(y;\mu)$.

Studentized unit deviance

Consider, for $\alpha > 0$, the new unit deviance

$$d_\alpha(y;\mu) = \frac{1}{\alpha}\log\{1 + \alpha d(y;\mu)\}, \tag{5.69}$$

called the **Studentized unit deviance** corresponding to $d(y;\mu)$. The limiting case $\alpha = 0$ corresponds to the original unit deviance d. The Studentized unit deviances give rise to the following class of regular proper dispersion models:

$$p(y;\mu,\lambda,\alpha) = c_\alpha(\lambda)V^{-\frac{1}{2}}(y)\exp\left\{-\frac{\lambda}{2}d_\alpha(y;\mu)\right\}, \tag{5.70}$$

denoted $\mathrm{PD}_\alpha(\mu,\sigma^2)$, with index set Λ_α. We denote the normalizing constant for the null model by $c(\lambda) = c_0(\lambda)$. We call $\mathrm{PD}_\alpha(\mu,\sigma^2)$ the **Student** or **Studentized** model corresponding to $\mathrm{PD}(\mu,\sigma^2)$.

The Studentized unit deviance $d_\alpha(y;\mu)$ satisfies the inequality

$$d_\alpha(y;\mu) \leq d(y;\mu),$$

which shows that the density (5.70) has heavier tails than the density of the null model $\mathrm{PD}(\mu,\sigma^2)$. In fact, the exponential factor of (5.70) decreases geometrically with $d(y;\mu)$ when $\alpha > 0$, but exponentially when $\alpha = 0$. The distribution (5.70) may hence be useful for outlier accommodation, providing a 'robust' alternative to the null model.

Mixing distribution

We now show that $\mathrm{PD}_\alpha(\mu,\sigma^2)$ may be interpreted as a Studentization of the null model $\mathrm{PD}(\mu,\sigma^2)$. Consider a random variable X with probability density function given by

$$q(x;\alpha,\lambda) = \bar{c}_\alpha(\lambda)\frac{x^{\frac{\lambda}{2\alpha}-1}}{\sqrt{2\pi}c(x)}\exp\left(-\frac{x}{2\alpha}\right) \tag{5.71}$$

for $x > 0$, where $\bar{c}_\alpha(\lambda)$ is a normalizing constant and $\alpha > 0$. Note that the integral

$$\frac{1}{\bar{c}_\alpha(\lambda)} = \int_0^\infty \frac{x^{\frac{\lambda}{2\alpha}-1}}{\sqrt{2\pi}c(x)}\exp\left(-\frac{x}{2\alpha}\right)dx \tag{5.72}$$

is always convergent at infinity, where $c(x) \sim \sqrt{x/2\pi}$. The domain for λ in (5.71) hence depends on the behaviour of c in zero. For example, if $c(x)$ behaves as $\sqrt{x}$ in zero, as is often the case, then λ has domain (α, ∞).

In particular, when $c(x) = \sqrt{x/2\pi}$, then X is a gamma random variable,
$$X \sim \text{Ga}\left(\lambda - \alpha, \frac{2\alpha}{\lambda - \alpha}\right).$$

The corresponding density is
$$q(x; \alpha, \lambda) = \bar{c}_\alpha(\lambda) x^{\frac{\lambda-\alpha}{2\alpha}-1} \exp\left(-\frac{x}{2\alpha}\right),$$
where now
$$\bar{c}_\alpha(\lambda) = \frac{1}{\Gamma\left(\frac{\lambda-\alpha}{2\alpha}\right)(2\alpha)^{\frac{\lambda-\alpha}{2\alpha}}}.$$

Mixture

Assume that $\Lambda_0 \supseteq \mathbf{R}_+$, and define the following mixture:
$$Y|X = x \sim \text{PD}\left(\mu, x^{-1}\right),$$
where X has density (5.71). Then the marginal density for Y is
$$\frac{\bar{c}_\alpha(\lambda)}{\sqrt{2\pi}} V^{-\frac{1}{2}}(y) \int_0^\infty x^{\frac{\lambda}{2\alpha}-1} \exp\left[-\frac{x}{2\alpha}\{1 + \alpha d(y; \mu)\}\right] dx$$
$$= \frac{\bar{c}_\alpha(\lambda) \Gamma\left(\frac{\lambda}{2\alpha}\right)(2\alpha)^{\frac{\lambda}{2\alpha}}}{\sqrt{2\pi}} V^{-\frac{1}{2}}(y) \exp\left\{-\frac{\lambda}{2} d_\alpha(y; \mu)\right\}. \quad (5.73)$$

Hence, comparing with the density (5.70), we find that Y has distribution $\text{PD}_\alpha(\mu, \sigma^2)$, and we obtain the following relation between the normalizing constants c_α and $\bar{c}_\alpha$:
$$c_\alpha(\lambda) = \frac{\bar{c}_\alpha(\lambda) \Gamma\left(\frac{\lambda}{2\alpha}\right)(2\alpha)^{\frac{\lambda}{2\alpha}}}{\sqrt{2\pi}}. \quad (5.74)$$

We thus have the following lemma.

Lemma 5.7 *The integral (5.72) is finite if and only if $\lambda \in \Lambda_\alpha$.*

In the special case where $c(x) = \sqrt{x/2\pi}$ we obtain
$$c_\alpha(\lambda) = \frac{\sqrt{\alpha}}{B\left(\frac{\lambda-\alpha}{2\alpha}, \frac{1}{2}\right)}. \quad (5.75)$$

The above mixture representation of the model $\text{PD}_\alpha(\mu, \sigma^2)$ provides a Studentization of the model $\text{PD}(\mu, \sigma^2)$, because the mixing distribution is either exactly or approximately a gamma distribution. This represents an analogy with the representation of the

Student t distribution as a normal mixture, see Section 5.4.4 below. The interpretation of the Studentized model as a mixture confirms the overdispersed nature of this model, as compared with the original model.

As mentioned above, we often have $\Lambda_\alpha = (\alpha, \infty)$, where $\alpha > 0$. Because of this constraint, the Studentization cannot in general be repeated on $\mathrm{PD}_\alpha(\mu, \sigma^2)$, due to the requirement $\Lambda_0 \supseteq \mathbf{R}_+$, while this may of course be possible in certain special cases, depending on the actual domain for λ.

Prediction of X

The conditional distribution of X given Y is a gamma distribution,

$$X|Y = y \sim \mathrm{Ga}\left\{\frac{\lambda}{1 + \alpha d(y;\mu)}, \frac{2\alpha}{\lambda}\right\}.$$

It is hence easy to predict the unobserved value of X from an observation of Y, which may be useful for diagnostic purposes if $\mathrm{PD}_\alpha(\mu, \sigma^2)$ is used for outlier accommodation, as mentioned above. We may think of the mixture model as having the dispersion σ^2 inflated by the factor $1 + \alpha d(y;\mu)$, compared with the null model.

5.4.2 From 'Student' to 'normal'

We now consider the inverse of the Studentization process above. Thus, let $\mathrm{PD}(\mu, \sigma^2)$ be a given regular proper dispersion model with unit deviance $d(y;\mu)$. Let $\mathrm{PD}_\beta(\mu, \sigma^2)$ denote the model corresponding to the transformed unit deviance

$$d_\beta(y;\mu) = \frac{1}{\beta}\left[\exp\{\beta d(y;\mu)\} - 1\right],$$

where $\beta > 0$, and let Λ_β denote the index set. When $\beta = \alpha$, this is the inverse of the transformation (5.69), by which we may hence reconstruct the 'normal' model from a given 'Student' model.

Proposition 5.8 *The index set Λ_β contains $\mathbf{R}_+$.*

Proof. We take $\alpha = \beta$, retracing our steps from above, in order to reconstruct the 'normal' model from the given 'Student' model $\mathrm{PD}(\mu, \sigma^2)$. Let c_β denote the normalizing constant corresponding to d_β,

$$\frac{1}{c_\beta(\lambda)} = \int_\Omega V^{-\frac{1}{2}}(y) \exp\left\{-\frac{1}{2}d_\beta(y;\mu)\right\} dy. \tag{5.76}$$

STUDENTIZATION

Now, consider the corresponding version of the integral (5.72),

$$\int_0^\infty \frac{x^{\frac{\lambda}{2\beta}-1}}{\sqrt{2\pi}c_\beta(x)} \exp\left(-\frac{x}{2\beta}\right) dx. \tag{5.77}$$

By Lemma 5.7, the integral (5.77) is finite if $\lambda \in \Lambda_0$. Hence, $1/c_\beta(x)$ is finite for all $x > 0$, implying that the integral (5.76) is finite for all $\lambda > 0$, and thereby concluding the proof. □

This result implies that the model $PD(\mu, \sigma^2)$ may be represented as a Studentization of the model $PD_\beta(\mu, \sigma^2)$. We call this the **deconvolution** of the model $PD(\mu, \sigma^2)$. Because of Proposition 5.8, the deconvolution may be repeated on the new model $PD_\beta(\mu, \sigma^2)$, and so on, creating a whole hierarchy of regular proper dispersion models. Each deconvolution step may be revoked by the corresponding Studentization step. As we have seen above, each Studentization step makes the tails of the density longer, and each deconvolution step makes the tails shorter. Since there is a limit to how long the tails can be, there is, in general, a limit to the number of Studentization steps one may take. This limit is reached once the Studentized model has an index set that does not include $\mathbf{R}_+$.

If we deconvolute after Studentization, but with $\alpha \neq \beta$, we obtain a transformed unit deviance of the form

$$\frac{1}{\beta}\left[\{1 + \alpha d(y; \mu)\}^{\beta/\alpha} - 1\right],$$

providing a further class of deviance transformations.

5.4.3 Exponentiation

We now consider briefly a third type of deviance transformations. Again, we let $PD(\mu, \sigma^2)$ denote a given regular proper dispersion model with unit deviance $d(y; \mu)$, and we assume that its index set Λ contains the value 0. This means that $V^{-1/2}(y)$ is integrable. Let $PD_\gamma(\mu, \sigma^2)$ denote the model corresponding to the transformed unit deviance

$$d_\gamma(y; \mu) = \frac{1}{\gamma}\left[1 - \exp\{-\gamma d(y; \mu)\}\right],$$

where $\gamma > 0$. We call this transformation **exponentiation**. In this case, the normalizing constant $c_\gamma(\lambda)$ may be calculated as follows:

$$\frac{1}{c_\gamma(\lambda)} = e^{-\frac{\lambda}{2\gamma}} \int_\Omega V^{-\frac{1}{2}}(y) \exp\left\{\frac{\lambda}{2\gamma} e^{-\gamma d(y;\mu)}\right\} dy$$

$$
\begin{aligned}
&= e^{-\frac{\lambda}{2\gamma}} \int_\Omega V^{-\frac{1}{2}}(y) \sum_{n=0}^{\infty} \frac{\lambda^n e^{-n\gamma d(y;\mu)}}{(2\gamma)^n n!}\, dy \\
&= e^{-\frac{\lambda}{2\gamma}} \sum_{n=0}^{\infty} \frac{\lambda^n}{(2\gamma)^n n!} \int_\Omega V^{-\frac{1}{2}}(y) e^{-n\gamma d(y;\mu)}\, dy \\
&= e^{-\frac{\lambda}{2\gamma}} \sum_{n=0}^{\infty} \frac{\lambda^n}{(2\gamma)^n n!\, c(2n\gamma)}. \quad (5.78)
\end{aligned}
$$

If we apply exponentiation to a Studentized model with parameter α, we obtain the following transformed unit deviance:

$$\frac{1}{\gamma}\left[1 - \{1 + \alpha d(y;\mu)\}^{-\gamma/\alpha}\right]. \quad (5.79)$$

The corresponding normalizing constant is given by (5.78) with $c(2n\gamma)$ replaced by $c_\alpha(2n\gamma)$, the normalizing constant from (5.70).

5.4.4 Examples

We now consider some examples of deviance transformations.

Student's t distribution

Studentization of the normal distribution gives Student's t distribution,

$$\frac{\sqrt{\alpha}}{B\left(\frac{\lambda-\alpha}{2\alpha}, \frac{1}{2}\right)}\left\{1 + \alpha(y-\mu)^2\right\}^{-\frac{\lambda}{2\alpha}},$$

with location μ, scale $1/\sqrt{\alpha}$ and degrees of freedom $\lambda/\alpha - 1$. The domain for λ is (α, ∞).

Studentized inverse Gaussian distribution

It is easy to see that the Studentized inverse Gaussian distribution is given by the following density:

$$\frac{\sqrt{\alpha}}{B\left(\frac{\lambda-\alpha}{2\alpha}, \frac{1}{2}\right)} y^{-\frac{3}{2}} \left\{1 + \alpha\frac{(y-\mu)^2}{y\mu^2}\right\}^{-\frac{\lambda}{2\alpha}}, \quad (5.80)$$

where y, μ and α are positive and $\lambda > \alpha$. By the transformation $y \mapsto 1/y$ and the reparametrization $\mu \mapsto 1/\mu$, (5.80) turns into

$$\frac{\sqrt{\alpha}}{B\left(\frac{\lambda-\alpha}{2\alpha}, \frac{1}{2}\right)} y^{-\frac{1}{2}} \left\{1 + \alpha\frac{(y-\mu)^2}{y}\right\}^{-\frac{\lambda}{2\alpha}}$$

STUDENTIZATION

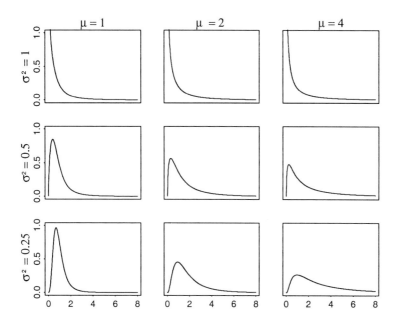

Figure 5.6. *Studentized inverse Gaussian densities with $\alpha = 1$.*

which is the Studentized version of the reciprocal inverse Gaussian distribution. Figures 5.6 and 5.7 show some examples of Studentized inverse Gaussian and reciprocal inverse Gaussian densities.

Studentized gamma distributions

For the gamma distribution we find, for x going to zero,

$$c(x) = \frac{x^x e^{-x}}{\Gamma(x)} = \frac{x^{x+1} e^{-x}}{\Gamma(x+1)} = O(x).$$

Hence, the Studentized gamma distribution has index set $\Lambda_\alpha = (2\alpha, \infty)$, and probability density function given by

$$c_\alpha(\lambda) y^{-1} \left\{ 1 + 2\alpha \left(\log \frac{\mu}{y} + \frac{y}{\mu} - 1 \right) \right\}^{-\frac{\lambda}{2\alpha}}.$$

It corresponds to the following mixing distribution:

$$q(x; \alpha, \lambda) = \bar{c}_\alpha(\lambda) \frac{x^{\frac{\lambda}{2\alpha}-1} \Gamma(x)}{x^x e^{-x}} \exp\left(-\frac{x}{2\alpha}\right). \quad (5.81)$$

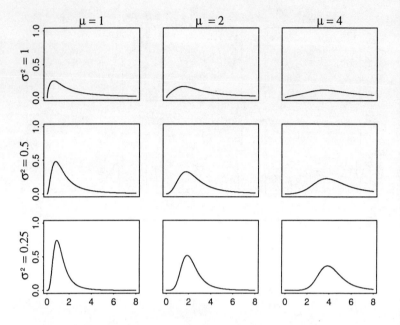

Figure 5.7. *Studentized reciprocal inverse Gaussian densities with $\alpha = 1$.*

Studentized simplex distributions

Consider the standard simplex distribution $S^-(\mu, \sigma^2)$, which has $c(\lambda) = \sqrt{\lambda/(2\pi)}$. The corresponding Studentized distribution is

$$\frac{\sqrt{\alpha}}{B\left(\frac{\lambda-\alpha}{2\alpha}, \frac{1}{2}\right)} \{y(1-y)\}^{-\frac{3}{2}} \left\{1 + \alpha \frac{(y-\mu)^2}{y(1-y)\mu^2(1-\mu)^2}\right\}^{-\frac{\lambda}{2\alpha}},$$

where $y, \mu \in (0,1)$ and $\lambda > \alpha$. Figure 5.8 shows some examples of Studentized standard simplex densities. Note that, for large σ^2, the densities are no longer unimodal, but go to infinity at 0 and 1.

The Studentized simplex distributions corresponding to the three other cases with $c(\lambda) = \sqrt{\lambda/(2\pi)}$ are easily derived in a similar manner. Table 5.2 summarizes the Studentized versions of the five simplex distributions from Table 5.1. The unit variance functions for the Studentized distributions are the same as for the corresponding distributions in Table 5.1.

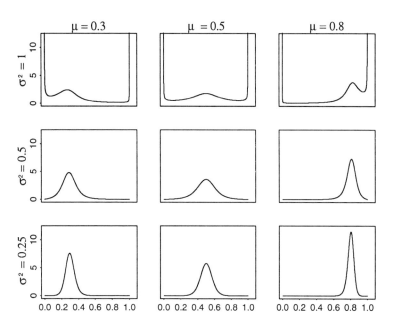

Figure 5.8. *Studentized standard simplex densities with $\alpha = 1$.*

Now, consider the case $(\alpha_1, \alpha_2) = (1/2, 1/2)$, corresponding to the unit deviance
$$d(y; \mu) = \frac{(y-\mu)^2}{y(1-y)}.$$

Table 5.2. *Studentized simplex distributions. The form of the normalizing constant in the case $(0,0)$ is not known*

(α_1, α_2)	$1/c_\alpha(\lambda)$	$d_\alpha(y;\mu)$
$\left(\frac{1}{2}, \frac{1}{2}\right)$	$B\left(\frac{\lambda+1}{2}, \frac{1}{2}\right)$	$\log\left\{1 + \frac{(y-\mu)^2}{y(1-y)}\right\}$
$(0,0)$	—	$\alpha^{-1}\log\left\{1 + \alpha\frac{(y-\mu)^2}{y(1-y)\mu(1-\mu)}\right\}$
$\left(-\frac{1}{2}, -\frac{1}{2}\right)$	$\alpha^{-\frac{1}{2}}B\left(\frac{\lambda-\alpha}{2\alpha}, \frac{1}{2}\right)$	$\alpha^{-1}\log\left\{1 + \alpha\frac{(y-\mu)^2}{y(1-y)\mu^2(1-\mu)^2}\right\}$
$\left(-\frac{1}{2}, \frac{1}{2}\right)$	$\alpha^{-\frac{1}{2}}B\left(\frac{\lambda-\alpha}{2\alpha}, \frac{1}{2}\right)$	$\alpha^{-1}\log\left\{1 + \alpha\frac{(y-\mu)^2}{y(1-y)\mu^2}\right\}$
$\left(\frac{1}{2}, -\frac{1}{2}\right)$	$\alpha^{-\frac{1}{2}}B\left(\frac{\lambda-\alpha}{2\alpha}, \frac{1}{2}\right)$	$\alpha^{-1}\log\left\{1 + \alpha\frac{(y-\mu)^2}{y(1-y)(1-\mu)^2}\right\}$

Recall that we proved, in Section 5.3.2, that this unit deviance is a proper unit deviance, because it is a function of the transformed Leipnik unit deviance. The Studentized version of this simplex distribution is

$$c_\alpha(\lambda)\left\{y(1-y)\right\}^{-\frac{1}{2}}\left\{1+\alpha\frac{(y-\mu)^2}{y(1-y)}\right\}^{-\frac{\lambda}{2\alpha}}, \qquad (5.82)$$

where again $y, \mu \in (0,1)$. The special case $\alpha = 1$ gives the transformed Leipnik distribution, and the corresponding normalizing constant is

$$c_1(\lambda) = \frac{1}{B\left(\frac{\lambda+1}{2}, \frac{1}{2}\right)}.$$

Since $\{y(1-y)\}^{-1/2}$ is integrable, the transformed Leipnik distribution may be used for exponentiation. In this case, the unit deviance (5.79) becomes

$$d_{\alpha,\gamma}(y;\mu) = \frac{1}{\gamma}\left[1 - \left\{1+\alpha\frac{(y-\mu)^2}{y(1-y)}\right\}^{-\gamma/\alpha}\right],$$

and the corresponding regular proper dispersion model becomes

$$c_{\alpha,\gamma}(\lambda)\left\{y(1-y)\right\}^{-\frac{1}{2}}\exp\left\{-\frac{\lambda}{2}d_{\alpha,\gamma}(y;\mu)\right\}.$$

In the special case $\alpha = 1$ the normalizing constant is

$$c_{1,\gamma}(\lambda) = e^{-\frac{\lambda}{2\gamma}}\sum_{n=0}^{\infty}\frac{\lambda^n}{(2\gamma)^n n! B\left(n\gamma+\frac{1}{2},\frac{1}{2}\right)}.$$

Studentized von Mises distributions

The Studentized von Mises distribution has the following form:

$$c_\alpha(\lambda)\left[1+2\alpha\left\{1-\cos(y-\mu)\right\}\right]^{-\frac{\lambda}{2\alpha}}, \qquad (5.83)$$

for $y, \mu \in (0, 2\pi)$, where $c_\alpha(\lambda)$ is a normalizing constant. In this case, the function $V^{-1/2}(y)$ is integrable, so the Studentized model may be exponentiated.

The exponentiated version of (5.83) corresponds to the following unit deviance:

$$d_{\alpha,\gamma}(y;\mu) = \frac{1}{\gamma}\left(1 - [1+2\alpha\{1-\cos(y-\mu)\}]^{-\gamma/\alpha}\right).$$

The corresponding model hence has the form

$$c_{\alpha,\gamma}(\lambda)\exp\left\{-\frac{\lambda}{2}d_{\alpha,\gamma}(y;\mu)\right\},$$

where the normalizing constant $c_{\alpha,\gamma}(\lambda)$ is given by

$$\frac{1}{c_{\alpha,\gamma}(\lambda)} = e^{-\frac{\lambda}{2\gamma}} \sum_{n=0}^{\infty} \frac{\lambda^n}{(2\gamma)^n n! \, c_\alpha(2n\gamma)}.$$

5.5 Notes

Most of the material in the present chapter is from Jørgensen (1997). As mentioned in Section 5.3.2, McCullagh's (1989) study of Leipnik's distribution provided a crucial incentive for the present study of proper dispersion models, by showing an example of a non-transformational proper dispersion model.

The (multivariate) simplex distribution was introduced by Barndorff-Nielsen and Jørgensen (1991).

As emphasized in Section 5.2.4, the distinction between transformational and non-transformational proper dispersion models is crucial for understanding the nature of proper dispersion models. Transformational dispersion models are easy to construct, and their structure is well understood, being derived from well-known principles for transformation models. By contrast, the very existence of non-transformational dispersion models, such as the simplex distribution or the transformed Leipnik distribution, seems quite mysterious mathematically. Statistically speaking, the question concerns the existence of pivots that are not related to transformation models.

One characterization of regular proper dispersion models is that their saddlepoint approximations are exact. Our understanding of non-transformational proper dispersion models is hence intimately related to our understanding of the properties of the saddlepoint approximation. In any case, we have already derived much benefit from studying the saddlepoint approximation, and, more generally, Barndorff-Nielsen's formula.

5.6 Exercises

Exercise 5.1 Consider a regular unit deviance $d(y;\mu)$ on $C \times \Omega$ with unit variance function V.

1. Show that d may be written in the form

$$d(y;\mu) = \frac{f(y;\mu)}{V(\mu)}(y-\mu)^2,$$

where f is strictly positive, continuous, and satisfies $f(y;y) = 1$ for $y \in \Omega$.

2. Assume that f factorizes as $f(y;\mu) = g(y)/g(\mu)$ for some function g. Show that if d is proper, then the corresponding regular proper dispersion model is an exponential family dispersion model.

3. Find f for the generalized inverse Gaussian and simplex unit deviances, and show that f factorizes as in Item 2. in both of these cases.

4. Calculate the six residuals discussed in Section 3.6, assuming the factorization in Item 2.

Exercise 5.2 Let Y_1, Y_2 be independent with distributions
$$Y_i \sim \text{PD}\left(\mu, \frac{\sigma^2}{w_i}\right), \quad i = 1, 2,$$
where $\text{PD}(\mu, \sigma^2)$ denotes a location-dispersion model. Let $\bar{Y} = \frac{1}{2}(Y_1 + Y_2)$ and $A = \frac{1}{2}(Y_1 - Y_2)$,

1. Show that
$$\bar{Y}|A = 0 \sim \text{PD}\left(\mu, \frac{\sigma^2}{w_1 + w_2}\right).$$

2. Comment on the relationship between this result and the convolution formula for exponential dispersion models.

Exercise 5.3 Let X follow Leipnik's distribution (5.50), and define
$$Y = \theta + \frac{(X - \theta)\left(\theta^2 - 1\right)}{1 - 2\theta X + \theta^2}.$$

1. Show that the probability density function of Y is given by
$$\frac{\left(1 - y^2\right)^{\frac{\lambda-1}{2}} \left(1 - \theta^2\right)}{B\left(\frac{\lambda+1}{2}, \frac{1}{2}\right)} \left(1 - 2y\theta + \theta^2\right)^{-\frac{\lambda}{2}-1},$$
where $-1 < y < 1$ (McCullagh, 1989).

2. Show that the above distribution for Y is not a proper dispersion model.

Exercise 5.4 Find the Studentized version of the hyperbola distribution.

Exercise 5.5 Consider the simplex distribution $S(1, -1, \mu, \sigma^2)$. Calculate the normalizing constant $c(1, -1, \mu, \lambda)$ for this distribution. Is this model a regular proper dispersion model?

EXERCISES

Exercise 5.6 Consider the family of probability density functions given by
$$p(y; \mu, \lambda) = c(\lambda) \exp\{\lambda \cos 2(y-\mu)\},$$
where $c(\lambda)$ is a normalizing constant, $y \in (0, 2\pi)$, $\mu \in (0, \pi)$ and $\lambda > 0$.

1. Find the normalizing constant $c(\lambda)$, and show that this is a transformational dispersion model.
2. Is the function $t(y; \mu) = \cos 2(y-\mu)$ yokable, and if so, is the model a regular proper dispersion model?

Exercise 5.7 Consider the unit deviance
$$d(y; \mu) = c + \log\{1 - \alpha \cos(y-\mu)\}, \quad y, \mu \in (0, 2\pi),$$
where $\alpha \leq 1$ and c are constants.

1. Find the value of c.
2. Show that d is proper.
3. Find the relationship, if any, between the model corresponding to d and the Studentized von Mises distribution.

Exercise 5.8 Derive the form of the six residuals discussed in Section 3.6 for a Studentized regular proper dispersion model.

Exercise 5.9 (Thinning) Let $\alpha \in (0,1)$ and let X be a random variable with distribution $\text{Tw}_p^*(\theta, \lambda)$, an additive Tweedie model. Define the distribution of the random variable Y by the following conditional distribution:
$$Y|X = x \sim \text{GTw}_p\{\alpha\lambda, (1-\alpha)\lambda, x\},$$
the Tweedie contraction from Section 5.3.5.

1. Show that Y and $X - Y$ are independent, $Y \sim \text{Tw}_p^*(\theta, \alpha\lambda)$ and $X - Y \sim \text{Tw}_p^*\{\theta, (1-\alpha)\lambda\}$. We call Y the **thinning** of X by the proportion α.
2. If $X \sim N(\lambda\theta, \lambda)$, show that $Y = \alpha X + X_0$, where
$$X_0 \sim N\{0, \alpha(1-\alpha)\lambda\}$$
is independent of X.
3. If $X \sim \text{Ga}^*(\theta, \lambda)$, show that $Y = GX$ where
$$G \sim \text{Beta}\{\alpha\lambda, (1-\alpha)\lambda\}$$
is independent of X.

References

Aalen, O.O. (1992). Modelling heterogeneity in survival analysis by the compound Poisson distribution. *Ann. Appl. Probab.* **2**, 951–972.

Abramowitz, M. and Stegun, I.A. (1972). *Handbook of Mathematical Functions.* New York: Dover.

Agresti, A. (1990). *Categorical Data Analysis.* New York: Wiley.

Bar-Lev, S.K. and Enis, P. (1986). Reproducibility and natural exponential families with power variance functions. *Ann. Statist.* **14**, 1507–1522.

Barndorff-Nielsen, O.E. (1977). Exponentially decreasing log-size distributions. *Proc. Roy. Soc.* A **353**, 401–419.

Barndorff-Nielsen, O.E. (1978a). *Information and Exponential Families in Statistical Theory.* Chichester: Wiley.

Barndorff-Nielsen, O.E. (1978b). Hyperbolic distributions and distributions on hyperbolae. *Scand. J. Statist.* **5**, 151–157.

Barndorff-Nielsen, O.E. (1980). Conditionality resolutions. *Biometrika* **67**, 293–310.

Barndorff-Nielsen, O.E. (1983). On a formula for the distribution of the maximum likelihood estimator. *Biometrika* **70**, 343–365.

Barndorff-Nielsen, O.E. (1986). Inference on full or partial parameters based on the standardized signed log likelihood ratio. *Biometrika* **73**, 307–322.

Barndorff-Nielsen, O.E. (1988). *Parametric Statistical Models and Likelihood.* Lecture Notes in Statistics Vol. 50. Berlin: Springer-Verlag.

Barndorff-Nielsen, O.E. (1989). Contribution to the discussion of R.E. Kass: The geometry of asymptotic inference. *Statist. Science* **4**, 222–227.

Barndorff-Nielsen, O.E. (1990). p^* and Laplace's method. *Braz. J. Probab. Statist.* **4**, 89–103.

Barndorff-Nielsen, O.E. and Blæsild, P. (1983). Reproductive exponential families. *Ann. Statist.* **11**, 770–782.

Barndorff-Nielsen, O.E. and Blæsild, P. (1988). Reproductive models. In *Encyclopedia of Statistical Sciences* Vol. 8 (eds S. Kotz and N.L. Johnson), pp. 86–89. New York: Wiley.

Barndorff-Nielsen, O.E., Blæsild, P., Jensen, J.L. and Jørgensen, B.

(1982). Exponential transformation models. *Proc. Roy. Soc.* A **379**, 41–65.

Barndorff-Nielsen, O.E. and Cox, D.R. (1979). Edgeworth and saddlepoint approximations with statistical applications (with discussion). *J. Roy. Statist. Soc. Ser.* B **41**, 279–312.

Barndorff-Nielsen, O.E. and Cox, D.R. (1989). *Asymptotic Techniques for Use in Statistics.* London: Chapman & Hall.

Barndorff-Nielsen, O.E. and Cox, D.R. (1994). *Inference and Asymptotics.* London: Chapman & Hall.

Barndorff-Nielsen, O.E. and Jørgensen, B. (1991). Some parametric models on the simplex. *J. Multivariate Anal.* **39**, 106–116.

Barndorff-Nielsen, O.E. and Klüppelberg, C. (1992). A note on the tail accuracy of the univariate saddlepoint approximation. *Ann. Fac. Sci. Univ. Toulouse*, Série 6 **1**, 5–14.

Berg, S. (1988). Stirling distributions. In *Encyclopedia of Statistical Sciences* Vol. 8 (eds S. Kotz and N.L. Johnson), pp. 773–776. New York: Wiley.

Bingham, N.H., Goldie, C.M. and Teugels, J.L. (1987). *Regular Variation.* Cambridge: Cambridge University Press.

Bishop, Y.M.M., Fienberg, S.E. and Holland, P.W. (1975). *Discrete Multivariate Analysis: Theory and Practice.* Cambridge, Massachusetts: MIT Press.

Blæsild, P. (1987). Yokes: Elemental properties with statistical applications. In *Geometrization of Statistical Theory* (ed. C.T.J. Dodson), pp. 193–196. Lancaster: ULDM Publications, Dept. Math., University of Lancaster.

Blæsild, P. (1991). Yokes and tensors derived from yokes. *Ann. Inst. Statist. Math.* **43**, 95–113.

Blæsild, P. and Jensen, J.L. (1985). Saddlepoint formulas for reproductive exponential models. *Scand. J. Statist.* **12**, 193–202.

Bondesson, L. (1990). Generalized gamma convolutions and complete monotonicity. *Probab. Theory Related Fields* **85**, 181–194.

Box, G.E.P. and Cox, D.R. (1964). An analysis of transformations. *J. Roy. Statist. Soc. Ser.* B **26**, 211–252.

Brown, L.D. (1986). *Fundamentals of Statistical Exponential Families With Application in Statistical Decision Theory.* Lecture Notes— Monograph Series, Vol. 9. Hayward, California: Institute of Mathematical Statistics.

Consul, P.C. and Jain, G.C. (1973). A generalization of the Poisson distribution. *Technometrics* **15**, 791–799.

Copson, E.T. (1965). *Asymptotic Expansions.* Cambridge: Cambridge University Press.

Cox, D.R. (1970). *The Analysis of Binary Data.* London: Chapman & Hall.

Critchlow, D.E. (1985). *Metric Methods for Analyzing Partially Ranked*

Data. Lecture Notes in Statistics Vol. 34. Berlin: Springer-Verlag.

Daniels, H.E. (1954). Saddlepoint approximations in statistics. *Ann. Math. Statist.* **25**, 631–650.

Daniels, H.E. (1980). Exact saddlepoint approximations. *Biometrika* **67**, 59–63.

Daniels, H.E. (1987). Tail probability approximations. *Internat. Statist. Rev.* **55**, 37–48.

de Haan, L. (1975). *On Regular Variation and its Applications to the Weak Convergence of Sample Extremes.* Mathematical Centre Tracts Vol. 32. Amsterdam: Mathematical Centre.

Eaton, M.L., Morris, C. and Rubin, H. (1971). On extreme stable laws and some applications. *J. Appl. Probab.* **8**, 794–801.

Efron, B. (1975). Defining curvature of a statistical problem (with applications to second order efficiency) (with discussion). *Ann. Statist.* **3**, 1189–1242.

Efron, B. (1986). Double exponential families and their use in generalized linear regression. *J. Amer. Statist. Assoc.* **81**, 709–721.

Feller, W. (1968). *An Introduction to Probability Theory and its Applications* Vol. I, Third Edition. New York: Wiley.

Feller, W. (1971). *An Introduction to Probability Theory and its Applications* Vol. II, Second Edition. New York: Wiley.

Fisher, R.A. (1934). Two new properties of mathematical likelihood. *Proc. Roy. Soc. Ser.* A **144**, 285–307.

Fisher, R.A. (1953). Dispersion on a sphere. *Proc. Roy. Soc. Ser.* A **217**, 295–305.

Haberman, S.J. (1974). *The Analysis of Frequency Data.* Chicago: University of Chicago Press.

Harkness, W.L. and Harkness, M.L. (1968). Generalized hyperbolic secant distributions. *J. Amer. Statist. Assoc.* **63**, 329–337.

Hoel, P.G., Port, S.C. and Stone, C.J. (1971). *Introduction to Probability Theory.* Boston: Houghton-Mifflin.

Holla, M.S. (1966). On a Poisson-inverse Gaussian distribution. *Metrika* **11**, 115–121.

Hougaard, P. (1986). Survival models for heterogeneous populations derived from stable distributions. *Biometrika* **73**, 387–396.

Hougaard, P. (1995). Nonlinear regression and curved exponential families. Improvement of the approximation to the asymptotic distribution. *Metrika* **42**, 191–202.

Hougaard, P., Lee, M.-L.T. and Whitmore, G.A. (1996). Analysis of overdispersed count data by mixtures of Poisson variables and Poisson processes. Unpublished report.

Jain, G.C. and Consul, P.C. (1971). A generalized negative binomial distribution. *SIAM J. Appl. Math.* **21**, 501–513.

Jenkins, G.M. (1956). Tests of hypotheses in the linear auto-regressive model. II Null distributions for higher order schemes: non-null distributions. *Biometrika* **43**, 186–199.

Jensen, J.L. (1981). On the hyperboloid distribution. *Scand. J. Statist.* **8**, 193–206.

Jensen, J.L. (1988). Uniform saddlepoint approximations. *Adv. Appl. Probab.* **20**, 622–634.

Jensen, J.L. (1989). Uniform saddlepoint approximations and log-concave densities. *J. Roy. Statist. Soc. Ser.* B **53**, 157–172.

Jensen, J.L. (1992). A note on a conjecture of H.E. Daniels. *Braz. J. Probab. Statist.* **6**, 85–95.

Johansen, S. (1979). *Introduction to the Theory of Regular Exponential Families.* Lecture Notes Vol. 3. Copenhagen: Institute of Mathematical Statistics, University of Copenhagen.

Johnson, N.L., Kotz, S. and Balakrishnan, N. (1994). *Continuous Univariate Distributions* Vol. I, Second Edition. New York: Wiley.

Johnson, N.L., Kotz, S. and Balakrishnan, N. (1995). *Continuous Univariate Distributions* Vol. II, Second Edition. New York: Wiley.

Johnson, N.L., Kotz, S. and Kemp, A.W. (1992). *Univariate Discrete Distributions*, Second Edition. New York: Wiley.

Jones, M.C. (1987). On the relationship between the Poisson-exponential model and the non-central chi-squared distribution. *Scand. Actuarial J.*, 104–109.

Jørgensen, B. (1982). *Statistical Properties of the Generalized Inverse Gaussian Distribution.* Lecture Notes in Statistics Vol. 9. New York: Springer-Verlag.

Jørgensen, B. (1983). Maximum likelihood estimation and large-sample inference for generalized linear and nonlinear regression models. *Biometrika* **70**, 19–28.

Jørgensen, B. (1984). The delta algorithm and GLIM. *Int. Statist. Rev.* **52**, 283–300.

Jørgensen, B. (1986). Some properties of exponential dispersion models. *Scand. J. Statist.* **13**, 187–198.

Jørgensen, B. (1987a). Exponential dispersion models (with discussion). *J. Roy. Statist. Soc. Ser.* B **49**, 127–162.

Jørgensen, B. (1987b). Small-dispersion asymptotics. *Braz. J. Probab. Statist.* **1**, 59–90.

Jørgensen, B. (1992). Exponential dispersion models and extensions: A review. *Internat. Statist. Rev.* **60**, 5–20.

Jørgensen, B. (1997). Proper dispersion models. *Braz. J. Probab. Statist.* (to appear).

Jørgensen, B. and Martínez, J.R. (1997). Tauber theory for infinitely divisible variance functions. *Bernoulli* (to appear).

Jørgensen, B., Martínez, J.R. and Tsao, M. (1994). Asymptotic behaviour of the variance function. *Scand. J. Statist.* **21**, 223–243.

Jørgensen, B. and Song, P.X.-K. (1998). Stationary time-series models with exponential dispersion model margins. *J. Appl. Probab.* **35** (to appear).

Kemp, A.W. (1970). General formulae for the central moments of certain serial correlation coefficient approximations. *Ann. Math. Statist.* **41**, 1363–1368.

Kemp, C.D. and Kemp, A.W. (1965). Some properties of the Hermite distribution. *Biometrika* **52**, 381–394.

Kokonendji, C.C. (1994). Exponential families with variance functions in $\sqrt{\Delta}P(\sqrt{\Delta})$: Seshadri's class. *Test* **3**, 123–172.

Kokonendji, C.C. and Seshadri, V. (1994). The Lindsay transform of natural exponential families. *Canadian J. Statist.* **22**, 259–272.

Küchler, U. (1982). Exponential families of Markov processes—Part I. General results. *Math. Operationsforsch. Statist., Ser. Statistics* **13**, 57–69.

Lamperti, J. (1966). *Probability. A Survey of the Mathematical Theory.* New York: Benjamin.

Lee, M.-L.T. and Whitmore, G.A. (1993). Stochastic processes directed by randomized time. *J. Appl. Probab.* **30**, 302–314.

Lehmann, E.L. (1983). *Theory of Point Estimation.* New York: Wiley.

Lehmann, E.L. (1986). *Testing Statistical Hypotheses*, Second Edition. New York: John Wiley.

Leipnik, R.B. (1947). Distribution of the serial correlation coefficient in a circularly correlated universe. *Ann. Math. Statist.* **18**, 80–87.

Letac, G. (1987). Discussion of B. Jørgensen: 'Exponential Dispersion Models'. *J. Roy. Statist. Soc. Ser. B* **49**, 154.

Letac, G. (1992). *Lectures on Natural Exponential Families and Their Variance Functions.* Monografias de Matemática Vol. 50. Rio de Janeiro: Instituto de Matemática Pura e Aplicada.

Letac, G. and Mora, M. (1990). Natural real exponential families with cubic variances. *Ann. Statist.* **18**, 1–37.

Lugannani, R. and Rice, S.O. (1980). Saddlepoint approximation for the distribution of the sum of independent random variables. *Adv. Appl. Probab.* **12**, 475–490.

Mallows, C.L. (1957). Non-null ranking models, I. *Biometrika* **44**, 114–130.

Matsunawa, T. (1986). Poisson distribution. In *Encyclopedia of Statistical Sciences* Vol. 7 (eds S. Kotz and N.L. Johnson), pp. 20–25. New York: Wiley.

McCullagh, P. (1983). Quasi-likelihood functions. *Ann. Statist.* **11**, 59–67.

McCullagh, P. (1989). Some statistical properties of a family of continuous univariate distributions. *J. Amer. Statist. Assoc.* **84**, 125–129.

McCullagh, P. (1993). Models on spheres and models for permutations. In *Probability Models and Statistical Analyses for Ranking Data* (eds M. Fligner and J.S. Verducci). Lecture Notes in Statistics Vol. 80, pp.

278–283. New York: Springer-Verlag.

McCullagh, P. and Nelder, J.A. (1989). *Generalized Linear Models*, Second Edition. London: Chapman & Hall.

Michael, J.R., Schucany, W.R. and Hass, R.W. (1976). Generating random variates using transformations with multiple roots. *Amer. Statistician* **30**, 88–90.

Mora, M. (1990). Convergence of the variance functions of natural exponential families. *Ann. Fac. Sci. Univ. Toulouse*, Série 5 **11**, 105–120.

Morris, C.N. (1981). *Models for positive data with good convolution properties*. Memo no. 8949. California: Rand Corporation.

Morris, C.N. (1982). Natural exponential families with quadratic variance functions. *Ann. Statist.* **10**, 65–80.

Murray, J.D. (1974). *Asymptotic analysis*. Oxford: Clarendon Press.

Nelder, J.A. and Pregibon, D. (1987). An extended quasi-likelihood function. *Biometrika* **74**, 221–232.

Nelder, J.A. and Wedderburn, R.W.M. (1972). Generalized linear models. *J. Roy. Statist. Soc. Ser.* A **135**, 370–384.

Pérez-Abreu, V. (1991). Poisson approximation to power series distributions. *Amer. Statistician* **45**, 42–45.

Pierce, D.A. and Schafer, D.W. (1986). Residuals in generalized linear models. *J. Amer. Statist. Assoc.* **81**, 977–986.

Read, C.B. (1985). Mill's ratio. In *Encyclopedia of Statistical Sciences* Vol. 5 (eds S. Kotz and N.L. Johnson), pp. 504–506. New York: Wiley.

Reid, N. (1988). Saddlepoint methods and statistical inference. *Statist. Science* **3**, 213–238.

Reid, N. (1995). The roles of conditioning in inference. *Statist. Science* **10**, 138–199.

Routledge, R. and Tsao, M. (1995). Uniform validity of saddlepoint expansion on compact sets. *Canad. J. Statist.* **23**, 425–431.

Rudin, W. (1976). *Principles of Mathematical Analysis,* Third Edition. New York: McGraw-Hill.

Sankharan, M. (1968). Mixtures by the inverse Gaussian distribution. *Sankhyā* **30**, 455–458.

Serfling, R.J. (1978). Some elementary results on Poisson approximation in a sequence of Bernoulli trials. *SIAM Review* **29**, 567–579.

Seshadri, V. (1994). *The Inverse Gaussian Distribution: A Case Study in Exponential Families*. Oxford: Clarendon Press.

Sichel, H.S. (1971). On a family of discrete distributions particularly suited to represent long-tailed frequency data. In *Proceedings of the Third Symposium on Mathematical Statistics* (ed. N.F. Laubscher). Pretoria: CSIR.

Siegel, A.F. (1985). Modelling data containing exact zeroes using zero degrees of freedom. *J. Roy. Statist. Soc. Ser.* B **47**, 267–271.

Sweeting, T.J. (1981). Scale parameters: A Bayesian treatment. *J. Roy. Statist. Soc. Ser.* B **43**, 333–338.

Sweeting, T.J. (1984). Approximate inference in location-scale regression models. *J. Amer. Statist. Assoc.* **79**, 847–852.

Tweedie, M.C.K. (1947). Functions of a statistical variate with given means, with special reference to Laplacian distributions. *Proc. Cambridge Phil. Soc.* **49**, 41–49.

Tweedie, M.C.K. (1957). Statistical properties of inverse Gaussian distributions, I. *Ann. Math. Statist.* **28**, 362–377.

Tweedie, M.C.K. (1984). An index which distinguishes between some important exponential families. In *Statistics: Applications and new directions. Proceedings of the Indian Statistical Institute Golden Jubilee International Conference* (eds J.K. Ghosh and J. Roy), pp. 579–604. Calcutta: Indian Statistical Institute.

Wasan, M.T. (1968). On an inverse Gaussian process. *Skandinavisk Aktuarietidsskrift* **60**, 69–96.

Wasan, M.T. (1969). *First Passage Time Distribution of Brownian Motion With Positive Drift (Inverse Gaussian Distribution)*. Queen's Papers in Pure and Applied Mathematics, No.19. Kingston, Ontario: Queen's University.

Wedderburn, R.W.M. (1974). Quasi-likelihood functions, generalized linear models and the Gauss-Newton method. *Biometrika* **61**, 439–447.

White, J.S. (1957). Approximate moments for the serial correlation coefficient. *Ann. Math. Statist.* **28**, 798–802.

Yanagimoto, T. (1989). The inverse binomial distribution as a statistical model. *Commun. Statist.* **18**, 3625–3633.

Symbol index

α 131, 193, 206
β 135, 161, 194, 208
γ 209
$\gamma_j(Y)$ 41
$\Gamma(\lambda)$ 18
$\Gamma(a,x)$ 200
ΔZ_n 83
θ 7, 42–43, 71–72, 108, 130–131, 176
Θ 43, 71, 176
Θ_p 131–132
Θ_Y 37
$\kappa(\theta)$ 42, 71
$\kappa_j(Y)$ 40
$\kappa_p(\theta)$ 131
λ 7, 72, 130, 175
Λ 71, 128, 175
μ 4, 46, 72, 127, 175
$\mu_j(Y)$ 40
ν 42, 71, 147, 151
ν_λ 71
$\bar{\nu}_\lambda$ 72
ξ 6–7, 25, 74, 84–85, 144
ρ 84, 144
σ^2 5, 72, 127–128, 175–176
$\tau(\theta)$ 46, 71, 73, 108
$\tau_p(\theta)$ 131
$\varphi_Y(t)$ 41
$\varphi(t;\theta,\lambda)$ 76
Φ 111
Φ^* 113
$\chi^2(f)$ 65, 123
Ω 4, 46, 71, 73
Ω_p 127

$a_0(\mu,\sigma^2)$ 27
$a(\sigma^2)$ 5
$a(y)$ 49
$a(y;\sigma^2)$ 5, 77
$a^*(z;\sigma^2)$ 6–7, 78

$b(y)$ 176
$B(\alpha,\beta)$ 33
$\mathrm{Bi}(m,\mu)$ 15

$c(\lambda)$ 176
$c_0(\mu,\lambda)$ 177
$c(y)$ 6, 43–44
$c(y;\lambda)$ 73, 175
$c(y;\mu,\lambda)$ 177
$c_p(y;\lambda)$ 141
$c^*(z;\lambda)$ 73
$c_p^*(z;\lambda)$ 137
C 4, 47, 45
C_λ 75

$\xrightarrow{d}$ 30
$d(y;\mu)$ 4, 49, 77
$d_0(y;\mu)$ 25
$d_p(y;\mu)$ 20, 134
$\mathrm{DM}(\mu,\sigma^2)$ 5, 175–176

$\mathrm{ED}(\mu,\sigma^2)$ 6, 72
$\mathrm{ED}^*(\theta,\lambda)$ 7, 72
$\mathrm{ED}^*(\Theta,\lambda)$ 72
$\mathrm{Ex}(\mu)$ 52

$F(y;\mu,\sigma^2)$ 111

SYMBOL INDEX

g 186
G 186
$\mathrm{Ga}^*(\theta,\lambda)$ 89
$\mathrm{Ga}(\mu,\sigma^2)$ 19, 88
$\mathrm{GHS}(\mu,\sigma^2)$ 102
$\mathrm{GIG}(\gamma,\chi,\psi)$ 158
$\mathrm{GTw}_p(\lambda_1,\lambda_2,x)$ 204

$I_0(\lambda)$ 21
$\mathrm{IG}(\mu,\sigma^2)$ 138
$\mathrm{IG}^*(\theta,\lambda)$ 140
int C 4

$K(\mu)$ 28
$K_0(\lambda)$ 34
$K_{\pm\frac{1}{2}}(\lambda)$ 201
$K_1(\lambda)$ 34
$K_\gamma(\omega)$ 158
$K_Y(s)$ 37
$K(s;\theta)$ 46
$K(s;\theta,\lambda)$ 132
$K_p(s;\theta,\lambda)$ 73
$K^*(s;\theta,\lambda)$ 73
$K_p^*(s;\theta,\lambda)$ 132

$M_Y(s)$ 37
$M(s;\theta)$ 44

$N(\mu,\sigma^2)$ 3
$\mathbf{N}$ 82
$\mathbf{N}_0$ 2
$\mathrm{Nb}(\lambda,p)$ 95
$\mathrm{NE}(\mu)$ 6, 46

$O(\sigma^2)$ 103

p 127
P_θ 43
$\mathcal{P}$ 43
$p(y;\theta)$ 44
$p(y;\theta,\lambda)$ 73
$p^*(z;\theta,\lambda)$ 73
$p^*(z;\xi,\sigma^2)$ 6, 78

$p(y;\mu,\lambda)$ 175
$p(y;\mu,\sigma^2)$ 5
$p_0(y;\mu,\sigma^2)$ 27
$p_0(y;\mu,\lambda)$ 182
$\mathrm{PD}(\mu,\sigma^2)$ 5, 175
$\mathrm{PD}_\alpha(\mu,\sigma^2)$ 206
$\mathrm{Po}(\mu)$ 14, 52

$q_Z(u)$ 64

r 110
r^* 111
r_P 108
r_W 109
$\mathbf{R}$ 2
$\mathbf{R}_+$ 2
$\mathbf{R}_-$ 38
$\mathbf{R}_0$ 2

s 109
S 2, 4, 47
$S^-(\mu,\sigma^2)$ 22
$S(\alpha_1,\alpha_2,\mu,\sigma^2)$ 199

$t(\nu)$ 33
$\hat{t}(y)$ 178
$t(y;\theta)$ 176
T 179
$\mathrm{TP}_p^*(\theta,\lambda)$ 167
$\mathrm{Tw}_p(\mu,\sigma^2)$ 127
$\mathrm{Tw}_p^*(\theta,\lambda)$ 130
$\mathrm{Tw}_\infty(\mu,\sigma^2,\beta)$ 161
$\mathrm{Tw}_\infty^*(\theta,\lambda,\beta)$ 161

u 109

$V(\mu)$ 4, 48, 73
$V_0(\mu)$ 26
$V_p(\mu)$ 127, 168
$\mathrm{vM}(\mu,\sigma^2)$ 21

Y 4, 37, 43, 127, 175

$\bar{Y}$ 11

Z 6, 72

$Z(t)$ 84, 144
Z 42
Z_+ 11

Author index

Aalen, O.O. 141, 219
Abramowitz, M. 21, 101, 191, 219
Agresti, A. 16, 219

Balakrishnan, N. 32, 196, 222
Bar-Lev, S.K. 170, 219
Barndorff-Nielsen, O.E. 5, 22, 28, 32, 48, 63, 104, 107, 111, 112, 113, 114, 116, 121, 158, 171, 181, 182, 183, 184, 187, 189, 192, 215, 219, 220
Berg, S. 124, 220
Bingham, N.H. 151, 220
Bishop, Y.M.M. 15, 220
Blæsild, P. 5, 187, 188, 189, 219, 220
Bondesson, L. 197, 200, 220
Box, G.E.P. 18, 35, 220
Brown, L.D. 63, 220

Consul, P.C. 155, 220, 221
Copson, E.T. 147, 220
Cox, D.R. 15, 18, 35, 104, 107, 113, 114, 116, 121, 220
Critchlow, D.E. 188, 220

Daniels, H.E. 32, 103, 104, 119, 121, 170, 188, 221
de Haan, L. 151, 154, 221

Eaton, M.L. 135, 164, 221
Efron, B. 27, 63, 221
Enis, P. 170, 219

Feller, W. 14, 37, 41, 42, 101, 135, 137, 150, 151, 221
Fienberg, S.E. 15, 220
Fisher, R.A. 1, 9, 11, 63, 221

Goldie, C.M. 151, 220

Haberman, S.J. 15, 221
Harkness, M.L. 101, 221
Harkness, W.L. 101, 221
Hass, R.W. 174, 180, 224
Hoel, P.G. 37, 221
Holla, M.S. 169, 221
Holland, P.W. 15, 220
Hougaard, P. 167, 170, 171, 221

Jain, G.C. 155, 220, 221
Jenkins, G.M. 195, 221
Jensen, J.L. 104, 112, 170, 187, 188, 192, 219, 220, 222
Johansen, S. 63, 222
Johnson, N.L. 32, 170, 196, 222
Jones, M.C. 170, 222
Jørgensen, B. 12, 20, 21, 22, 31, 32, 58, 59, 63, 69, 94, 120, 147, 152, 155, 158, 159, 170, 187, 193, 204, 215, 219, 220, 222

Kemp, A.W. 32, 123, 170, 195, 222, 223
Kemp, C.D. 123, 223
Klüppelberg, C. 171, 220
Kokonendji, C.C. 120, 157, 223

AUTHOR INDEX

Kotz, S. 32, 170, 196, 222
Küchler, U. 120, 223

Lamperti, J. 84, 223
Lauritzen, S.L. 180
Lee, M.-L.T. 167, 170, 221, 223
Lehmann, E.L. 63, 223
Leipnik, R.B. 195, 196, 197, 198, 214, 215, 216, 223
Letac, G. 120, 156, 157, 158, 169, 223
Lugannani, R. 112, 113, 114, 121, 126, 223

Mallows, C.L. 188, 223
Martínez, J.R. 58, 59, 63, 69, 147, 152, 155, 159, 170, 222
Matsunawa, T. 63, 223
McCullagh, P. 12, 107, 110, 170, 180, 188, 195, 196, 215, 216, 223, 224
Michael, J.R. 174, 180, 224
Mora, M. 31, 54, 55, 57, 58, 59, 63, 79, 82, 120, 149, 151, 156, 165, 223, 224
Morris, C.N. 31, 63, 85, 91, 98, 100, 101, 102, 120, 126, 135, 156, 164, 170, 174, 221, 224
Murray, J.D. 147, 224

Nelder, J.A. 2, 12, 15, 18, 20, 27, 107, 110, 120, 224

Pérez-Abreu, V. 59, 224
Pierce, D.A. 110, 224
Port, S.C. 37, 221

Pregibon, D. 27, 224

Read, C.B. 113, 224
Reid, N. 32, 95, 111, 121, 224
Rice, S.O. 112, 113, 114, 121, 126, 223
Routledge, R. 104, 224
Rubin, H. 135, 164, 221
Rudin, W. 54, 224

Sankharan, M. 169, 224
Schafer, D.W. 110, 224
Schucany, W.R. 174, 180, 224
Serfling, R.J. 63, 224
Seshadri, V. 120, 171, 223, 224
Sichel, H.S. 169, 224
Siegel, A.F. 170, 224
Song, P.X.-K. 204, 222
Stegun, I.A. 21, 101, 191, 219
Stone, C.J. 37, 221
Sweeting, T.J. 17, 31, 224–225

Teugels, J.L. 151, 220
Tsao, M. 58, 59, 63, 69, 104, 147, 152, 159, 170, 222, 224
Tweedie, M.C.K. 31, 63, 119, 120, 128, 170, 171, 225

Wasan, M.T. 145, 225
Wedderburn, R.W.M. 2, 15, 18, 20, 63, 120, 224, 225
White, J.S. 195, 225
Whitmore, G.A. 167, 170, 221, 223

Yanagimoto, T. 155, 225

Subject index

Additive model 72
Additive process 83–85
  continuous time 84
  discrete time 83
Additive semigroup 82
Analysis of deviance 2, 12
Analysis of variance 2

Barndorff-Nielsen's p^* formula 32, 181–184
  exactness of 182–184, 187
Bernoulli distribution 59, 92, 94, 95, 97, 98–99
Bernoulli process 93–94
Bessel function 21, 34, 121–122, 158, 191, 201
Beta-binomial distribution 201, 203
Beta distribution 22, 33, 67, 195, 196, 199, 201, 204, 217
Beta function 33
Binomial data 2, 201–204
Binomial distribution 91, 126, 196, 204
  additive form 15, 92–94
  natural exponential family form 53
  Poisson approximation 59
  unit deviance 86
  variance-stabilizing transformation 35
Binomial mixture 15
Bondesson class 197–198, 200
Box-Cox transformation 35

Brownian motion 88, 145, 195

Canonical parameter 43, 72
  generalization of 108
Canonical parameter domain 43, 71
Canonical statistic 43
Cauchy distribution 38, 39, 64, 122
Central Limit Theorem 14, 55–56, 63, 124, 150
Central limit type 149–150, 165, 174
Characteristic function 41–42, 76–77
  continuity theorem 42
  inversion formula 41–42
  uniqueness theorem 41
Chi-square distribution 65, 123, 172
Compositional data, *see* Proportions
Compound Poisson distribution 20, 129, 140–144
Compound Poisson-Poisson mixture 169–170
Compound Poisson process 146
Continuous proportions, *see* Proportions
Convergence in distribution 30
Convergence of variance functions 54–57

Convergence to
  compound Poisson
    distribution 157
  exponential distribution
    56–57, 156
  gamma distribution 79, 97,
    103, 126, 150, 152, 156, 157,
    158
  inverse Gaussian distribution
    155, 156, 158, 174
  normal distribution 14, 30, 33,
    35, 56, 78–79, 89, 92, 94, 96,
    123, 124, 160, 165, 174, 181
  Poisson distribution 59, 63, 69,
    79–80, 94, 96, 123, 124, 155,
    156, 157
  Tweedie model 148–149,
    159–160, 164–165
Convergence type 149–150
Convex function 38, 64, 68
Convex support 4, 47, 75, 175
Convolution formula 80–81
  additive form 80
  reproductive form 81–82
Count data 2, 14–15, 76
Cumulant 40–41, 65
  standardized 41, 65
Cumulant function 42–43
  domain of 43
  unit 71–72
Cumulant generating function
  37–39
  continuity theorem 42
  convexity of 39
  degenerate 38
  domain of 37, 39, 64
  proper 40
  uniqueness theorem 41
Cumulant generator, see
  Cumulant function

Data on the real line 2
Deconvolution 209
Degenerate distribution 38
Deviance 2, 4, 49–50, 68
  example of nonregular 5
  proper 185
  regular 4, 24–25, 175, 215–216
  reparametrization of 25, 110
  Studentized 206
  total 12–13
  transformed 180–181, 205
  unit 4, 10–11, 24, 32, 49, 77
Directed deviance, see Residual,
  deviance
Directions 2, 20–21
Discrete data, see Count data
Discrete uniform distribution 68
Dispersion 1, 2, 9, 12
Dispersion model 2, 3, 5, 10
  additive form 6–7, 11, 15, 72,
    78
  continuous 75–76
  discrete 75–76
  exponential family 176, 194
  general 176, 177–179
  non-transformational 7, 8, 190
  probability density function 5,
    6, 9
  reproductive form 5, 11, 72, 77
  standard form 5, 175, 178
  transformational 7, 8, 186–188
  type 7
  see also Exponential
    dispersion model; Proper
    dispersion model
Dispersion parameter 5, 9–10, 72
  interpretation 19–20
Divisible 82, 150
Division/convolution 98
Double exponential distribution,
  see Laplace distribution
Double exponential family 27
Duality transformation 6, 72, 76

Effective domain 37
Expectation 40, 46
Exponential dispersion model 6,
  71–78
  additive form 7, 11, 72

SUBJECT INDEX 233

continuous 75–76, 85
convex support 75
cumulant generating function 73
cumulants 122
discrete 75–76
mean 73–74
parametrizations 73–75
probability density function 73, 77–78
reproductive form 6, 11, 72
unit deviance 77
unit variance function 73, 77
variance 73–74
Exponential distribution 19, 52, 56
cumulants 66
moment generating function 37–38
Exponential family 63, 199
reproductive 188
see also Natural exponential family
Exponential family dispersion model 176
Exponential power family 34
unit deviance 5
Exponential tilting 43
Exponentiated Leipnik distribution 214
Exponentiated von Mises distribution 214–215
Exponentiation 209–210
Extended quasi-likelihood 27
Extreme stable distribution 135

Gamma distribution 8, 18–20, 27, 85, 88–90, 123, 127–128, 171, 188, 207, 217
residuals 114–115
saddlepoint approximation 104–105
unit deviance 86
variance-stabilizing transformation 35

Gamma process 90, 145
General dispersion model, see Dispersion model, general
Generalized hyperbolic secant distribution 85–86, 100–103, 126
Generalized inverse Gaussian distribution 158–160, 186, 190–194, 200
Generalized linear model 2, 12–13, 15–16
Generalized Poisson distribution 155–156, 173
Geometric distribution 66
GLIM 16

Hermite distribution 122–123
Hölder's inequality 38
Hougaard process 170
Hyperbola distribution 34, 192
Hyperbolic distribution 34, 121
Hyperbolic secant distribution 101

Incomplete gamma function 200, 201
Index for stable distribution, see Stable distribution
Index parameter 71–72
Index set 72, 77, 82–83, 175–176, 196
Infinite divisibility 82–84, 123, 150, 197
Infinitely divisible type 149–150, 165, 174
Inverse binomial distribution 155
Inverse Gaussian distribution 8, 119, 137–140, 171, 172, 174, 188, 191
Inverse Gaussian-Poisson mixture 168–169
Inverse Gaussian process 145, 174

Karamata's Theorem 154

Kolmogorov's Consistency
   Theorem 84
Kurtosis 41, 122

Laplace approximation 28
Laplace distribution 17, 68
Lattice distribution 75–76
Leipnik's distribution 176,
   195–197, 214, 215, 216
   transformed 196–197, 198
Length biased sampling 67
Letac form 157–158
Lévy process 84
Locally $\text{Tw}_p(\mu, \sigma^2)$ 150
Location-dispersion model 17, 34,
   186, 187
Location parameter 3, 9–19
Location-scale model 17, 162
Log likelihood 4, 5, 10
Logarithmic distribution 67, 124
Logistic distribution 69
Lugannani–Rice formula 112–113,
   126, 172–173

Maximum likelihood estimator 4,
   10, 12, 32, 49, 182
Mean domain 46, 71, 73
Mean value mapping 46, 71, 73
   inverse 108
Mean value parameter 46, 72
Mills' ratio 113
Mode point 9
Moment 40–41
Moment generating function
   37–39
   continuity theorem 42
   degenerate 38
   domain 37, 64
   proper 40
   uniqueness theorem 41
Mora's convergence theorem
   54–55, 56, 58–59, 63, 82–83,
   149, 151, 165
   see also Convergence to
Morris class 85, 97–100, 126

Natural exponential family 6, 13,
   42–45, 71–72
   continuous case 44
   cumulant generating function
      46
   cumulants 46, 65–66
   discrete case 44
   example of non-steep 68
   generated by distribution 43,
      65
   generated by measure 43
   mixed case 44
   moment generating function
      44
   moments 46
   probability density function 44
   regular 48
   steep 48
Negative binomial distribution
   15, 16, 91, 124, 126, 168
   additive form 94–97
   unit deviance 86
Negative binomial process 97
Neyman type A distribution 170
Noncentral chi-square
   distribution 170
Normal distribution 1, 2, 6, 8, 17,
   26, 98, 127–128, 171, 188, 217
   exponential dispersion model
      form 85–88
   natural exponential family
      form 65
   mixture 174
   moment and cumulant
      generating function 64
'Normal' model 206
Normalizing constant 177
Normed yoke 32
Null model 205

Order 146–148
Overdispersion 167, 203, 206, 208

p^* formula, see
   Barndorff-Nielsen's p^* formula

SUBJECT INDEX   235

Pivot 179–181, 215
Poisson distribution 14–15, 59,
    74–75, 98, 124, 127–128, 192
  additive form 16, 75, 90–92
  natural exponential family
    form 52–53
  reproductive form 127, 171
  residuals 119–120
  standardized cumulants 66
  unit deviance 86
  variance-stabilizing
    transformation 35
Poisson mixture 166–167
Poisson process 91, 146
Position-dispersion model 10
Position parameter 2, 5, 10
  interpretation 19–20
Positive data 2, 18–20
Positive data with zeros 2, 20
Positive stable distribution 135
Power series distribution 59
Probability generating function
    64–65, 122
Proper dispersion model 5, 29,
    179–181
  construction 184–186
  exponential family
  general 176
  non-transformational
  probability density function 5,
    176
  regular 5–6, 175
  reproductive form 5, 176
  standard form 5, 176
  transformational 186–187
Proportions 2, 22–24
Pseudo-dispersion model 177,
    181, 205
Pure jump process 90

Quadratic variance function, see
  Variance function, quadratic

Ranking model 188
Rare event 14–15

Realizable values 4, 175
Reciprocal gamma distribution
    158, 194
Reciprocal inverse Gaussian
    distribution 191
Regular variation 151
Renormalized saddlepoint
    approximation 27
Reparametrization 25
Reproductive model 11, 82
Residual 108–111, 217
  approximate normality 111
  crude 109
  deviance 110–111
  dual score 109
  modified deviance 111
  Pearson 108
  score 109
  standard 110
  Wald 109

Saddlepoint approximation 26–29,
    32, 121, 171, 173, 204–205
  continuous case 103–106
  discrete case 106–108
  exactness 26, 105, 189, 215
  renormalized 27–29, 35, 177,
    184–185, 189
  uniform on compacts 28, 104
Scale-dispersion model 18, 194
Scale parameter 3, 9–10
Scaled Poisson distribution 66, 74
Simplex-binomial mixture
    201–204
Simplex distribution 22–23, 26,
    198–204, 215, 216
  residuals 115–116
  standard 22–24, 198–199, 204
Skewness 41, 111, 122
Slow variation 151
Small dispersion 11
Squared distance 4, 9, 12, 26
Stability index 135
Stable distribution 135, 139, 158,
    163

Stable Tweedie model 136
Standard form 5
Standardized cumulant 41
Steepness 48
Stirling distribution 124
Stirling's formula 105, 123
'Student' model 206
Studentization 205–209
   mixing distribution 206
   null model 205
   prediction 208
   residuals 217
   Studentized model 206
Studentized gamma distribution 211
Studentized hyperbola distribution 216
Studentized inverse Gaussian distribution 210–211
Studentized reciprocal inverse Gaussian distribution 210–211
Studentized simplex distributions 212–214
Studentized von Mises distribution 214–215, 217
Student's t distribution 33, 210
Support 4, 5, 43, 47, 58, 65

t distribution, see Student's t distribution
Tail area approximation 111–113, 121
Tauber Theorem 151
Tauber type 149–150, 151–155, 174
Thinning 217
Transformation 110
   affine 65, 98, 121
   of dispersion model 25
   scale 128
   translation 160, 187
   of unit deviance 180–181
   variance-stabilizing 26, 29, 35
Transformation invariance 110
Transformational dispersion model, see Dispersion model, transformational
Truncated Poisson distribution 66
Tweedie contraction 204–205
Tweedie convergence theorem 148–149
Tweedie model 127–134, 161, 170
   additive form 130, 161, 172
   canonical parameter domain 132
   characterization of 128
   convolution formula 162–163
   cumulant function 131
   cumulant generating function 132
   location transformation 161
   mean domain 128
   mean value mapping 131
   probability density function 137, 141–142
   reproductive form 127, 161, 172
   residuals 142–144
   scale transformation 129, 161
   stable 135–137
   unit deviance 133–134
   unit variance function 128
Tweedie-Poisson mixture 167–169
Tweedie process 144–146
Type 149

Uniform distribution 57, 67, 123
Uniform on compacts 28
Unit cumulant function, see Cumulant function
Unit deviance, see Deviance
Unit variance function, see Variance function

Variance 40, 48
Variance function 4, 24, 48–49, 50, 68, 175
   asymptotic behaviour 57–63, 69

SUBJECT INDEX

convergence 54–57, 69, 78–80, 149, 165
cubic 120, 156, 174
exponential 160–162, 170, 173
Letac form 157–158, 169
power 127, 162
quadratic 97–100
regularity 146–148, 164, 168, 170
uniqueness theorem 50–52, 63, 74, 100
unit 4, 10–11, 24, 73, 77

Variance-stabilizing transformation 26
von Mises distribution 21, 187
residuals 116–118

World map of dispersion models 8–9

Yokable function 178–179, 187
Yoke 5, 32, 49
normed 32